Philosophy of Biology

PRINCETON FOUNDATIONS OF CONTEMPORARY PHILOSOPHY

Scott Soames, *Series Editor*

Philosophical Logic by JOHN P. BURGESS
Philosophy of Language by SCOTT SOAMES
Philosophy of Law by ANDREI MARMOR
Truth by ALEXIS G. BURGESS and JOHN P. BURGESS
Philosophy of Physics: Space and Time by TIM MAUDLIN
Philosophy of Biology by PETER GODFREY-SMITH

PHILOSOPHY
OF BIOLOGY

Peter Godfrey-Smith

PRINCETON UNIVERSITY PRESS
PRINCETON AND OXFORD

Copyright © 2014 by Princeton University Press
Published by Princeton University Press,
  41 William Street, Princeton, New Jersey 08540
In the United Kingdom: Princeton University Press,
  6 Oxford Street, Woodstock, Oxfordshire OX20 1TW

press.princeton.edu

LIBRARY OF CONGRESS CATALOGING-IN-PUBLICATION DATA
Godfrey-Smith, Peter.
Philosophy of biology / Peter Godfrey-Smith.
pages cm
Includes bibliographical references and index.
ISBN 978-0-691-14001-8 (cloth : acid-free paper) 1. Biology—
  Philosophy. I. Title.
QH331.G615 2013
570—dc23   2013024519

British Library Cataloging-in-Publication Data is available

This book has been composed in Minion and Archer

Printed on acid-free paper ∞

Printed in the United States of America

10 9 8 7 6 5 4 3 2 1

For David Hull (1935–2010)

Contents

Preface

THIS BOOK WAS written primarily for readers of two kinds: philosophy students and biologists interested in philosophical issues surrounding their work. Writing for both these audiences leads to a fair amount of stage setting on both the philosophical and biological sides, and I hope this will make the book accessible to people without either of those backgrounds as well.

References relevant to the main line of argument are in the text and footnotes. The "Further Reading" sections at the end of each chapter mostly list works that take different approaches to the issues of that chapter, along with surveys and collections, though I sometimes repeat a reference to a work that was especially important in that chapter.

In writing this book I have benefitted greatly from contributions made by students in classes at Stanford University, Harvard University, and the City University of New York. For comments on an entire earlier draft I am grateful to Marc Ereshefsky, Arnon Levy, John Matthewson, Thomas Pradeu, Jane Sheldon, Derek Skillings, Elliott Sober, Kim Sterelny, Michael Weisberg's research group at the University of Pennsylvania, Kritika Yegnashankaran, and an anonymous reader for the press. Sterelny once again had a significant impact on the transition between near-final and final versions. For comments that helped me with particular chapters and issues I am grateful to Guillaume Beaulac, Karen Bennett, Austin Booth, Herb Gintis, Andreas Keller, Ben Kerr, Enoch Lambert, David McCandlish, and Adam See. Brett Calcott and Eliza Jewett-Hall assisted with the figures. Thanks to Rob Tempio at Princeton for his patience and for his help at several stages.

The book is dedicated to the late David Hull, a pioneer in this field and a continuing inspiration.

New York City
March 2013

Philosophy of Biology

Philosophy and Biology

In working out how philosophy and biology are related, and what the philosophy *of* biology might be, much depends on general questions about the nature of philosophy and what it aims to achieve. The best one-sentence summary of what philosophy is up to was given by Wilfrid Sellars in 1962: philosophy is concerned with "how things in the broadest possible sense of the term hang together in the broadest possible sense of the term." Philosophy aims at an overall picture of what the world is like and how we fit into it.

Science, too, tries to work out how things "hang together." Philosophy does this in an especially broad way, but breadth comes in degrees. As a result, some philosophical work shades off into science; there is not a sharp border between them. Philosophy also shades off into fields like politics, law, and mathematics. In its relation to science, philosophy has often also functioned as an "incubator" of theoretical ideas, a place where they can be developed in a speculative way while they are in a form that cannot be tested empirically. Many theories seen now in psychology and linguistics, for example, have their origins in philosophy. I think of this incubator role as secondary, though, and as separate from the role that the Sellars quote expresses.

1.1. What is the philosophy of biology?

Given this picture of philosophy, what relation does philosophy have to biology? One part of the totality that "hangs together" somehow, as Sellars put it, is the world of living things, like ourselves, other animals, plants, and bacteria. Another part of the totality is human *investigation* of the living world, including the practice of science. Here are some examples of philosophical issues that arise in and around biology, in roughly the order in which they appear in this book.

Although modern biology seems to have given us a good understanding of the living world, it seems to have done so without, for the most part, describing that world in terms of *laws*, as many sciences do. Is this because the subject matter of biology is special, because the science is less advanced, or because there are plenty of laws of biology but we are not calling them by that name? That is one of the topics of the second chapter, which also looks at the role of "mechanistic" explanations in biology and at the role of theoretical models that seem to roam far from actuality, even though they aim to help us understand the empirical world.

The book then turns to evolution, and the third chapter focuses on the most controversial part of evolutionary theory, Darwin's idea of *natural selection*. Many puzzles arise around what exactly can be explained in terms of selection, and in terms of the associated idea of biological "fitness." The last part of the chapter looks at the application of evolutionary ideas outside the usual borders of biology; Darwinian ideas have been applied to change in practices within a culture, for example, and to ideas jostling around in a person's head. Are these applications of Darwinian thinking just loose metaphors, or is change by natural selection a universal feature of biological, social, and psychological systems?

One of the most historically influential and psychologically powerful ways of thinking about living things is in terms of their *purposes* and *functions*. Modern biology, with its combination of a mechanistic, bottom-up treatment of biological processes and an evolutionary account of how living things come to be, has an uneasy relationship with that way of thinking. Does this package of views dissolve the appearance of purpose in the biological world, or explain where purposes come from? This is one topic of the fourth chapter, which also looks at some elusive questions about the relationships organisms have to their circumstances of life: to what extent do organisms *adapt* to their environments, and to what extent do they *construct* them? The fifth chapter is about organisms themselves, and other "individuals" in biology. It looks at what sort of things these are, how they are bounded, and how

they come to exist. The sixth is about genetics. It begins by looking at the changing role of genes as objects, as hidden factors that
explain what organisms are like. I then turn to their role in evolution, especially the idea that all of evolution can be seen as a long-
term struggle between rival genes.

The seventh chapter discusses species and other biological
kinds. Are species real units, objective aspects of the living world's
structure, perhaps with "essences" that mark off one kind of organism from another? Chapter 8 is about social behavior, and it
looks closely at *cooperation* and related phenomena. I outline a
general theory of the evolution of cooperative behaviors, a theory that takes a very abstract form, and then turn to the special
case of cooperation in human societies. How much similarity is
there between cooperation as a human, psychologically complex
phenomenon and cooperation or coordination between the un-
thinking parts of living systems? After this discussion of social
behavior I look at how the discussions of species in chapter 7 and
social behavior in chapter 8 fit together to tell us something about
"human nature," if such a thing exists at all.

The last chapter looks at another social phenomenon that has
deep roots running through living systems: communication. This
topic connects to a larger debate about the role of *information* in
biology. Some biologists think that evolutionary processes, perhaps life itself, are in some sense *made* of information. I discuss
those ideas fairly critically, but then look at recent work on the
ways that signaling and communication pervade living systems,
and at models of the evolution of these special forms of interaction. How does information transmission of the sort that we are
engaged in now, reading and writing, relate to what goes on inside
our bodies, between genes and cells?

These are some of the themes the book will look at. With this
list laid out, it is possible to see a further way of organizing things,
and thinking about the role of philosophy in relation to biology.
In some of the areas described above the goal of the philosopher
is to understand something about *science*—how a particular
part of science works. In other cases, the goal is to understand
something about the natural world itself, the world that science is
studying.

In a broad sense, all philosophy of biology is part of the "philosophy of science." But with an eye to the distinction just made, we can also distinguish *philosophy of science*, in a narrower sense, from *philosophy of nature*. Philosophy of science in this narrower sense is an attempt to understand the activity and the products of science itself. When doing philosophy of nature, we are trying to understand the universe and our place in it. The science of biology becomes an instrument—a lens—through which we look at the natural world. Science is then a resource for philosophy rather than a subject matter.

Though science is a resource for the philosopher trying to understand life, philosophy has its own perspective and its own questions. It is foolish for philosophy to place itself *above* science, but it can certainly step back from science and gain an outsider's viewpoint. This is necessary, in fact, for philosophy to be able to pursue the task of seeing how *everything* hangs together. A philosopher will look at how the message of one part of science relates to that of another, and how the scientific view of nature relates to ideas we get from other sources. The philosopher's vantage point makes it natural to question things that might be taken for granted, perhaps for practical reasons, within scientific work. So the project I call "philosophy of nature" is not giving a philosophical *report* of what is going on in science, but working out what the raw science is really telling us, and using it to put together an overall picture of the world.

This is not something that only philosophers can do. Scientists often have their own views about the philosophical significance of their work, and we'll encounter these views often in this book. But distilling the philosophical upshot of scientific work is a different activity from doing science itself.

The activity of science is itself part of nature; it is an activity undertaken by human agents. These two kinds of philosophical work interact; what you think science is *telling* us about the world will depend on how you think that part of science *works*. But being interested in the activity of science and being interested in what science is telling us about the world are somewhat different things, both of them part of the view of philosophy expressed by Sellars in the quote at the start of this chapter.

1.2. Biology and its history

This section gives a brief historical sketch of some parts of biology, emphasizing the development of evolutionary ideas and general views of the living world. The aim is to introduce some of the biological theories that are important in the book, including both current ideas and older ones that provide context and contrasts. A later chapter includes a separate historical survey of genetics.

Many early theories about the living world included evolutionary speculations of some kind—ideas about how familiar living things might have their origins in other kinds of life, or in nonliving matter. Among the ancient Greeks, Empedocles (ca. 490–430 BC) is an interesting example. He held that the earth had given birth to living creatures, but these first creatures had been disembodied *parts* of familiar organisms: "arms wandered without shoulders, and eyes strayed in need of foreheads."[1] These parts joined into combinations, with some surviving and others, unfit for life, disappearing. So the organisms we see now are results of a simple kind of "selection" process. Variations appeared and some were kept while others were lost.

Plato and Aristotle, the most influential ancient philosophers, did not endorse an evolutionary picture. In Aristotle's work a different kind of change, the orderly progression within each lifetime from egg to adult, was observed carefully and seen as a paradigm of "natural" and goal-directed change. He also saw movement towards goals as central to understanding change in areas far from biology, including physical phenomena. Living things for Aristotle are connected by gradations, with a scale from lower to higher forms that connects plants, animals, and man, though this scale does not reflect a historical sequence. The idea of a scale between higher and lower, a *scala naturae*, was immensely influential in the centuries to follow, forming an important part of the fusion of Aristotle's philosophy with Christianity that guided thinking through the Middle Ages. These scales typically began in inanimate things, extended through plants to simple and complex animals, then to man, the angels, and God.

---

[1] This is from Aristotle's account of Empedocles in *On Nature*.

As knowledge of plants and animals improved, scales from higher to lower came to seem less and less adequate. Some writers began to represent the organization of life with branching trees, along with other more complicated shapes (O'Hara 1991). They generally did not think of these trees and other shapes as representing patterns of ancestry. They were thought to represent "affinities"—similarities in underlying form—which have a basis in the "plan of the Creator." In the mid-18th century Carl Linnaeus developed the system of classification that is still used—in modified form and with some controversy—today (Linnaeus 1758). This is a system of groups within groups. Linnaeus categorized organisms initially in terms of their *kingdom, class, order, genus,* and *species.* (Other categories, such as *phylum* and *family,* were added later.)

Evolutionary speculation continued to crop up. The 18th-century French naturalist Buffon wondered about the common ancestry of some species. Darwin's grandfather Erasmus proposed in *Zoonomia* (1794) that all life diverged from a primordial "filament." The suggestion that new forms might appear by chance, some flourishing and others dying off, was sketched in vague form by various writers. The French enlightenment philosopher Denis Diderot included the idea in an anonymously published antireligious pamphlet that was so controversial that when Diderot was found to be the author he was thrown in jail ("Letter on the Blind," 1749).

The first detailed evolutionary theory was developed by Jean-Baptiste Lamarck, working in the early 19th century in France. Lamarck is famous now for the idea that evolution can occur by the "inheritance of acquired characteristics," something often referred to as "Lamarckian" evolution. The idea is that if an organism acquires a new physical characteristic during its lifetime, as a consequence of its habits of life, there is some tendency for that characteristic to be passed to its offspring. A hypothesis that Lamarck put more emphasis on, however, involved the actions of fluids, visible and invisible, flowing through living bodies. As they flow, they carve out new channels and make each organism more complex, in a way inherited across generations (Lamarck 1809). Life for Lamarck is also continually produced from inanimate matter by "spontaneous generation," forming independent lineages. A mammal alive now, for Lamarck, is a member of an older evolutionary lineage than a jellyfish around now; the jellyfish lineage has had less time to travel the road toward increased complexity. The present mammal and jellyfish do not have a common ancestor, though the mammal has a long-dead jellyfish ancestor. Lamarck did use a tree-like drawing to represent the relations between groups of organisms. There is some debate about how it should be interpreted, but it was not a tree representing a total pattern of common ancestry.[2]

Charles Darwin worked out his central ideas in the 1830s and published *On the Origin of Species* in 1859, publishing then because another English biologist, Alfred Russell Wallace, had come to similar conclusions. Darwin's theory had two main parts. One was a hypothesis of *common ancestry* of living species, which Darwin presented in terms of a "tree of life." As noted above, tree metaphors had been used to represent the organization of life before this. Darwin's move was to give the tree a historical, genealogical interpretation. Through evolutionary time, new species are formed by the splitting or fragmentation of existing ones. This gives rise to a network of relatedness among species themselves, forming the shape of a tree.

The other part of Darwin's view was a theory of how change occurs within species—on twigs or segments of the tree. In any species, new variations appear from time to time by accident. Individuals appear with quirks in their structure or behavior that other members of the species do not have. These variations arise in a haphazard way (perhaps, according to Darwin, due to shocks to the reproductive system). Most new variations are harmful, but a few help organisms to survive and reproduce. Many of these characteristics also tend to be passed on in reproduction. When a new characteristic appears that both is useful *and* tends to be inherited, it is likely to proliferate through the species. Small

---

[2] A comment Lamarck made in defense of this view has considerable evolutionary irony. He noted that a version of his view exists as a proverb, "*Habits form a second nature.*" Then, "if the habits and nature of each animal could never vary, the proverb would have been false and would not have come into existence, nor been preserved in the event of anyone suggesting it" (1809/2011, p. 114).

changes of this kind accumulate, and slowly give rise to whole new forms of life.

Darwin's thinking was influenced by three sets of ideas in other fields. "Natural theology" was a tradition of writing about nature emphasizing the perfection of God's creation, especially the complex design of organisms and the match between organism and environment (Paley 1802/2006). A second influence was Thomas Malthus's *Essay on the Principle of Population* (1798), a pessimistic work that argued that the natural growth of the human population must inevitably lead to famine, as the food supply could never grow fast enough to keep up. This led Darwin to the idea of a "struggle for life." The third was Charles Lyell's work in geology (1830), which argued that dramatic transformations of the earth could result from the operation of undramatic, everyday causes operating over vast periods of time.

Darwin was cautious on many points. He was unsure whether life formed a single tree or several. He accepted that factors beside natural selection affect the evolutionary process. He did not tie his view to speculations about matters about which little was known, such as the physical nature of life—he avoided the "fluids" and "filaments" of earlier writers. Instead he linked his evolutionary hypotheses to familiar and readily observed phenomena, especially the results of animal and plant breeding.[3]

Most biologists were fairly quickly convinced that evolution (as we now call it) had occurred, and that common ancestry connects much or all of life on earth. There was more controversy about *how* the process had happened, especially about natural selection and Darwin's insistence on gradual change. One of the weaker points in Darwin's work was his understanding of reproduction and inheritance. Gregor Mendel, a monk working in what is now the Czech Republic, had worked out some crucial ideas in this area around 1860, but his work was largely ignored. Mendel suggested that inheritance is due to "factors" (later called

[3] A remark in a letter by William James in 1883 captures, in James's unique style, an aspect of Darwin's mind that made his work so powerful: Darwin's tendency was to avoid abstractions and consider "concrete things in the plenitude of their peculiarities & with all the consequences thereof" (Skrupskelis 2007, p. 747).

"genes") that are passed on intact across generations, forming new combinations in different individuals. In 1900 this work was rediscovered and the science of genetics emerged. Initially, many scientists thought that the new Mendelian ideas were incompatible with Darwinism, as the Mendelian view was seen as allied to a "discontinuous" or "saltationist" view of evolution in which new forms appear in sudden jumps.

In time, Darwin's ideas were united with Mendelian genetics (Fisher 1930, Wright 1932). According to this "synthesis" of the views, most characteristics of organisms are affected by many genes, each of which has small effects. Evolution occurs as selection and other factors gradually make genes more or less common in the "gene pool" of the species. New genes are introduced by the random "mutation" of old genes. So mutation produces new genes, sexual reproduction brings existing genes into new combinations, and natural selection makes genes more or less common, as a result of the overall effect each gene has on the construction of living organisms.

One thing missing from this picture was any understanding of the chemical makeup of genes, and the processes by which they affect organisms. Another problem was the absence of much connection between evolutionary theory and the biology of individual *development*; evolution, according to critics, was being presented as if comprised of a procession of adults. The first changed in 1953, with the discovery of the double-helix structure of DNA by James Watson and Francis Crick. This discovery contained immediate clues about *how* genes do what they do (Crick 1958). The years that followed saw a deluge of information from the new "molecular biology," adding a further level of detail to evolutionary theory as the rest of biology was transformed.

In the past few pages I followed evolutionary thinking from the early 19th century forward. Central ideas in other parts of biology were also established in the 19th century. These include the ideas that *cells* are the basic units in living things, and that cells arise from other cells by division and fusion. Experiments by Louis Pasteur put the idea of ongoing "spontaneous generation" of life to rest in the middle of the century. For many years the chemistry of living systems, or "organic" chemistry, had seemed

9

so separate from the rest of chemistry that it appeared that life might involve its own special chemical principles, beyond those seen in "inorganic" matter. This also changed in the 19th century, with the first chemical synthesis of organic compounds and recognition of the special role of carbon, with its ability to form complex structures such as rings and chains. The puzzlingly separate "organic" chemistry became carbon chemistry.

Nonetheless, debate continued through the late 19th and early 20th centuries over whether all living activity has a purely physical basis. "Vitalists" thought that living processes were too purpose-driven to be merely physical (Driesch 1914). The biology of individual development, the sequence by which egg leads to adult, remained so puzzling that for some it did seem possible that a special organizing factor, something beyond ordinary physics, might be operating. Vitalism faded as the mechanistic side of biology advanced, and late in the 20th century the orderly progression that Aristotle had seen as a paradigm of natural change received a new type of explanation through the integration of developmental biology with molecular genetics, and a charting of the intricate processes by which gene action is regulated within cells. Simultaneously, the effects on evolutionary paths of the processes of individual development were explored (especially by the "evo-devo" movement), integrating explanations of change from the levels of molecules, through organisms, to the evolution of species.

FURTHER READING

For large-scale history, see Lovejoy (1936), Bowler (2009); for Lamarck, Burkhardt (1977); for Darwin, Browne (1996, 2003) and Lewens (2006); for precursors, including those outside the Western tradition, Stott (2012); on the synthesis, Provine (1971), J. Huxley (1942); on evolution and development, Amundson (2005), Laublichler and Maienschein (2009), Wagner (forthcoming); on species, Wilkins (2009); on molecular biology, Judson (1996).

Laws, Mechanisms, and Models

LOOKING AT BIOLOGY from a philosophical point of view, one of the first things people notice is that there is apparently not much role for scientific *laws*. The image of science as a search for the laws governing the natural world is an old and influential one, and many philosophers have held that the investigation of laws is central to any genuine scientific field (Carnap 1966, Hempel 1966). The laws of physics may be basic, but each science tries to find its own laws—laws present in the systems it studies. Perhaps biology is just a cataloguing of the world's contents, and not a theoretical science that gives us real understanding?[1] The progress in biology over the past century has made this seem more and more unlikely. Instead, it appears that good science can be organized differently. Or perhaps laws are present in biology but we are not seeing them clearly and calling them by that name?

This chapter is about the organization of hypotheses and explanations in biology. I start with laws, and then look at two other sets of issues.

2.1. LAWS

What exactly is a law of nature? There is much disagreement, and I will focus on a few features that are widely accepted. First, a statement of a law is a true generalization that is *spatiotemporally unrestricted*; it applies to all of space and time. Second, a law does not describe how things merely *happen* to be, but (in some sense) how they *have* to be. An example of a law that seems to

---

[1] Ernest Rutherford, who split the atom, allegedly said, "All science is either physics or stamp collecting." See also Smart (1959).

meet these criteria is Einstein's principle that no signal can travel faster than light. The idea that laws describe how things must, or have to, be is sometimes expressed by saying that laws have a kind of *necessity*. That may seem an overly strong word, and some philosophers would avoid it. Laws are not supposed to have the same kind of necessity seen in mathematical or logical truths (such as "*p&q* implies *p*"). But even if the term "necessity" is not used, there is supposed to be a distinction between a natural law and an "accidental" regularity, a pattern that merely happens to hold. A standard example of an accidental regularity that might be true for all space and time is: *all spheres of gold have a diameter of less than one mile*. Contrast: *all spheres of uranium-235 have a diameter of less than one mile*. A sphere of uranium that big would explode, so this second regularity is one that *has* to hold.

Laws might be "strict"—admitting no exceptions—or they might involve probabilities. That divide will not matter much here. There is also a verbal ambiguity: sometimes the term "law" is used for a *statement* of one of these patterns in nature, and sometimes for the pattern itself. I will use the term for the patterns themselves.

A biological example that has been much discussed is "Mendel's First Law." This principle has been revised since the days of early genetics, and it has exceptions. But it is a good illustration of several aspects of the situation. In modern language, the principle says that in the formation of sex cells (eggs and sperm), a diploid organism (one with two sets of chromosomes, like us) puts one gene into each sex cell of the two genes that it received at that place in its genome from its own parents, and each of these two genes has a 50 percent probability of being found in any given sex cell. Exceptions include cases of Down syndrome in humans, and cases where particular genes have evolved the capacity to make their way into more than their fair share of sex cells (§6.3). But let's set those aside for now and treat the generalization as near enough to true in sexually reproducing species like ours. Is this a "law," or an "accidental" regularity?

The best initial answer seems to be "a bit of *both*." There is no reason to think that any sexually reproducing animal *must* do things this way. The genetic system we find in organisms like us evolved from something different, and might evolve into something else in the future (Beatty 1995). You might say that the generalization holds uniformly within organisms on earth over some relevant period, but laws are not supposed to be restricted to some places and times.

On the other hand, if we look at sexual organisms that are around now, it is *no accident* that the regularity holds. Once certain machinery is in place, this machinery has consequences, and these include the patterns described in Mendelian genetics. Mendel's First Law, to the extent that it holds, is a predictable result of the operation of mechanisms that are contingent historical products.

I'll put another couple of examples on the table. The "Central Dogma of Molecular Biology" describes the construction of new protein and nucleic acid molecules. It holds that the specification of the order of the building blocks in these molecules always goes *from* nucleic acid *to* protein, never vice versa and never from protein to protein (Crick 1958, 1970). (The Central Dogma is sometimes described as saying more than that, but I will stick with Crick's version.) "Kleiber's Law" describes the rates at which animals of different sizes use energy. The metabolic rate of an animal (R) depends on its body mass (M) and a constant (c) according to this formula: $R = cM^{3/4}$. Discovered in the 1930s, this relationship holds across a wide range of cases. For different groups of animals there is a slightly different c (so c is a "constant" only within each group) but the ¾ is always the same. For many years this was seen as a striking and mysterious relation, and then it turned out to be possible to derive Kleiber's Law from general features of the transport networks that move substances around the body, such as blood vessels, along with an assumption that efficiency in these networks is maximized. Given those assumptions, Kleiber's Law must hold (West et al. 1997).

Kleiber's Law initially seems independent of history, a manifestation of general facts about transport networks. But what about the assumption that these networks will be efficiently organized? Biologists differ on how unusual it is for evolution to produce inefficient or poorly adapted outcomes, but it is certainly possible.

13

In both the Mendel case and the Kleiber case, "law or accident?" seems to be the wrong question to ask. So let's start again. Rather than a two-way distinction between laws and accidental regularities, biological patterns show different amounts of what can be called *resilience* or *stability*. (Other terms used in this area are *robustness* and *invariance*.)² A resilient pattern is one that holds across many actual cases, and does so in a way that gives us reason to believe it would also hold in some relevant situations that are not actual but merely possible. A resilient pattern need not hold in *any* possible situation, and it might have some actual-world exceptions. Resilience is not a yes-or-no matter; regularities have more or less of it, roughly speaking, though there isn't a single scale on which all can be compared.

Mendel's First Law and Kleiber's Law have some degree of resilience, though in both cases it is clear how exceptions could arise. Let's look again at the Central Dogma. When we look at what the Dogma rules *out*, it looks quite resilient. An exception would be some process in which the order of the amino acids (the building blocks) in a protein molecule was used to determine the order of nucleotides in a molecule of DNA or RNA, or the amino acids in another protein. This is thought to be chemically difficult. We might be wrong in thinking this, and perhaps an actual-world exception will be found. But so far the Central Dogma looks pretty resilient. On the other hand, the thing that made the Dogma important—the thing that made it reasonable to use the term "Central"—was the idea that proteins are made by simply reading off the sequence of nucleotides in DNA. Complications to that picture, including the discovery of widespread "editing" of the RNA intermediate stages, have steadily grown (§6.1), and as they have grown the centrality of the Dogma has shrunk.

So far I have been discussing broad and well-known principles. Biology also has many narrower generalizations with some degree of resilience. In mammals, the sex of an individual is determined by its male parent (except perhaps in one enigmatic vole). Spiders

---

²I borrow this term from Skyrms (1980), one of the first to introduce an idea of this kind, but I use the term differently. See Woodward (2001) for differences within this family.

are carnivorous. (For years I used this as an example of a pattern without any exceptions, but now a vegetarian has been found [Meehan et al. 2009].) Some generalizations in biology describe the ways that biological properties are distributed among actual organisms. Others describe the causal consequences of a setup or interaction of factors, without saying where or how often this setup is found: a species that has lost almost all of its genetic diversity is likely to go extinct. That principle describes the consequences of low diversity without saying which species fall into that category.

Once these facts have all been laid out, you might decide to use the term "law" for all the patterns that have *some* resilience, you might reserve it for cases that have a great deal, or you might think the term should be dropped from biology. That is mostly a verbal choice.

Does this analysis apply to *all* of science, or just to biology and similar fields? Sandra Mitchell (2000) applies a view of this kind to all of science, including physics.[3] Another possibility is that physics is a special case; physics describes laws that govern the fundamental working of the world, and the working of these laws in organisms gives rise to further patterns that are not much like physical laws but have various degrees of resilience.

2.2. MECHANISMS

I'll say more about laws later, but first I will look at some newer accounts of how theories work in biology. One family of views hold that large parts of biology are engaged in the *analysis of mechanisms.*[4] A mechanism is an arrangement of parts that produces a more complex set of effects in a whole system in a regular way. Biology describes how DNA replication works, how photosynthesis works, how the firing of one neuron makes another fire. In cases like these, the activities of the parts of a system are described, and

[3] As Mitchell notes about the standard example on the second page of this chapter, given the way gold comes to exist in a universe like ours it is *not* so "accidental" that huge amounts of it have never come together in one place.

[4] Central works here are Bechtel and Richardson (1993), Glennan (1996), and Machamer et al. (2000).

these activities and the relations between them explain how the more complicated capacities of the whole system arise.

To say this is not yet to break from a law-based view. Perhaps the way mechanisms are analyzed is by showing how the parts are governed by laws? But this approach is often seen a replacement for a law-based view. In the analysis of mechanisms, a different kind of causal understanding seems to be sought, or at least available. This is visible in the language used to describe causal relationships: one molecule will *bind* to another, altering how it interacts with other molecules. Or it might *cleave* or *oxidize* another. A stretch of DNA will be *transcribed*, or *silenced* by the *methylation* of some of its sequence. This seems to be a form of causal description oriented around the idea that some events *produce* others, in virtue of how things are physically connected. (I will discuss causal relations again, and modify this picture, in §6.2.) Generalizations expressed in these terms might still be seen as describing laws, but laws don't play an overt role in this sort of analysis. And although the parts of a neuron firing or embryo developing may well be following *physical* laws, that is in the background, and there seems to be no need to find laws of biology if you can describe all the mechanisms in this way.

This kind of work is "reductionist," in a low-key sense of that term: the properties of whole systems are explained in terms of the properties of their parts, and how those parts are put together. Reductionism is sometimes associated with the idea that a whole system is "nothing but" its lower-level parts, but this "nothing but" talk is usually quite misleading. A living system may be entirely composed out of a collection of parts, but the system will have features that none of the parts have. Rather than showing that the higher-level activities do not exist, the point of mechanistic explanation is usually showing *how* the higher-level features arise from the parts.

This view gives a good account of at least part of biology. How far does it extend? One option is to extend it very broadly. Perhaps natural selection, for example, is a mechanism in this sense, and evolutionary biology is about the analysis of mechanisms?[5]

[5] See Skipper and Millstein (2005). The next few pages have been influenced by Levy (2013) and Matthewson and Calcott (2011).

Perhaps the exclusion of one species by another in an ecological system is also a mechanism? I think saying those things requires diluting the sense of "mechanism" that has been useful in the analyses sketched above. Instead, there is a side of biology that analyzes mechanisms and a side that does not.

The philosophers who argue for the importance of mechanistic analysis do not tie their view to a 17th-century sense of the term "mechanism," in which the universe is treated as if it were clockwork. Mechanistic views of that kind, which see the world as governed only by pushes, pulls, and collisions, have been rejected in basic physics. But the biological systems to which mechanistic analysis most directly applies do have a machine-like quality in another sense. They are not only physical systems, but *organized* ones. This is another vague term, but a way to make sense of it is to think about how sensitive a system is to small changes in its parts, especially substitutions of one part for another. If we look at a neuron firing or a protein being made, the process we are interested in occurs as a result of the interactions of parts whose exact relations to each other matter. If you swapped a chromosome for a ribosome, the consequences would usually be large, just as they would be if you swapped a car's fuel pump for its gear box. One part of biology is concerned with systems like this. Other areas are concerned with systems that are "looser." When a population of organisms is evolving, there are parts (the organisms) and a whole (the population), but many of the parts are similar to each other, and if you swapped one for another it often would not make that much difference. Their exact relationships—who is next to whom—do not matter so much. These relationships *might* matter greatly in a particular case, but often they do not. A gas, as studied in physics, is a more extreme case of the same thing. A gas contains many molecules moving about in specific ways, but the details of those ways do not matter to various important properties of the gas, like its temperature and pressure. Those properties depend on broad and general features of the collection, such as the molecules' average speed. If you swap one molecule for another, it usually won't matter. When analyzing a system of that kind, a statistical approach is often taken.

Fields like evolutionary biology, ecology, and epidemiology are concerned with systems of this second kind—or more exactly,

17

not with *un*organized systems but with *less* organized systems. The systems they study are more organized than a gas, but less organized than a cell.

Adapting some terminology used by Richard Levins (1970) and William Wimsatt (2006, 2007), we can distinguish more *organized* systems from more *aggregative* ones. More organized systems include cells and organisms, and more aggregative ones include populations of those organisms. There are intermediate cases, like honey bee colonies. Sometimes if you "zoom out" from an aggregative system, organization will reappear. If you imagine watching gas molecules interacting with blood cells in the lungs of a large animal, what you see will be an aggregative system. If you swap one oxygen molecule for another, it does not make much difference. But if you zoom out so that the lungs become one organ in a whole body, *that* system is a highly organized one. Mechanistic analysis is most appropriate when dealing with organized systems. Aggregative ones are better described in terms of tendencies that arise from the combined action of parts which each have some degree of independence. The two kinds of systems occur at different scales; there are size constraints, it seems, on highly organized systems, due to the difficulty of keeping the parts working together, and organized systems often have distinctive kinds of histories. Organized systems often make use of aggregative activities in their small parts (consider molecules diffusing across a membrane). It is interesting to think about human societies, which can be very large objects, in terms of this distinction.

This distinction can also be used to clear up, or perhaps replace, another. Earlier I mentioned the idea of "reduction." A term often used to express a contrast with features that can be reductively explained is "emergent." Emergent properties are sometimes said to be those that *can't be explained* at a lower level; they are "irreducible." In the philosophy of mind, consciousness is sometimes said to be an emergent property in this sense. The claim is that although consciousness has some material basis in the brain, it can't be explained at the neural level. You could know exactly what all the neurons were doing, and it would still be mysterious why those brain processes gave rise to consciousness.

The term "emergent" is also used in much weaker (more inclusive) ways. Biologists sometimes use it to refer to properties of a whole system that the system's individual parts do not have. The high-level properties might be *explained* in terms of the parts, but are not *present* at the lower level. An example is the "surface tension" phenomenon in water. Surface tension is a consequence of the tendency of water molecules at an air/water boundary to form lots of weak chemical bonds with each other rather than with the air. An individual water molecule does not have surface tension; the phenomenon exists only when many molecules are brought together. This is a sense of "emergent" in which most features of any complex system will qualify.

The underlying phenomenon here is, once again, something like a gradient: higher-level activities in a system can be more or less dependent on the exact relations between the parts. If you want to draw a line between the "emergent" and the "reducible" properties, you could draw it at the divide between cases where higher-level properties are also *present* in the parts and cases where they are not, but then emergent properties are often clearly explainable in a bottom-up way. The distinctions beyond that one are distinctions of degree. How sensitive is a high-level behavior—the music coming from an orchestra, the economic patterns coming from the choices of individuals, the behavior coming from a collection of human cells—to the arrangement of the system's parts, in addition to the parts' individual properties? The idea of a special category of emergent properties that cannot be explained at all in lower-level terms has been influenced by the special perplexities of the mind/body problem. It probably does not help there, and there is no support for such a picture in other parts of biology.

2.3. MODELS

Of the other styles of work in biology, one is especially relevant here. This is modeling, or model building.

"Model" has many meanings in science and philosophy. Sometimes the term is used to describe any theory or hypothesis, or to describe a theory that is acknowledged to be rough or simplified.

However, the word can also be used to indicate a particular strategy in scientific work, a strategy in which one system is used as some kind of surrogate for another. The usual reason for doing this is that the "target" system we want to understand is too complicated to investigate directly. So it makes sense to choose some of the most important factors operating in the target system, and work out how they interact in a situation in which other factors are absent. Alternatively, one system might be used as a model for another because the best available methods can be more easily applied to the model than to the target, even though the model system is no simpler.[6]

In some cases a "model system" will be a physically built object. Engineers still build scale models of river systems and bays. This is related to the use of "model organisms" in biology. Model organisms, such as fruit flies and E. coli bacteria, were initially naturally occurring organisms that were easy to work with in the lab. Now they are often partly artificial, with features that would never occur in the wild. Much modeling work is different from this, however, as there is no model system that is physically present. Instead the model system is imagined or hypothetical. A researcher will write down a set of assumptions that are relevantly similar to those that hold in some real system, and will use mathematical analysis, computer simulation, or some other method to work out the consequences of those assumptions.

Evolutionary game theory is an example of a field where this method is widespread. Game theory uses mathematics to study how rational agents should behave in relation to each other. In the 1970s George Price and John Maynard Smith pioneered the use of this method to deal with animal behavior.[7] Rather than assuming that animals are rational, they assumed that natural selection will lead to the proliferation of behaviors that promote survival and reproduction, eliminating behaviors that do not. The first application of these methods was to fighting; the aim was to work out why bluffing and ritualized non-damaging fights are so common in animals. Here is one result. Suppose we have a popu-

---

[6] See Giere (1988), Godfrey-Smith (2006), Weisberg (2007b, 2013).
[7] Maynard Smith and Price (1973), Maynard Smith (1982).

lation in which individuals meet at random, one on one, and fight over resources. The population contains two kinds of individuals, "hawks" who will fight until they win or are seriously injured, and "doves" who bluff initially but retreat if things get out of hand. Individuals who do well in these contests are assumed to reproduce more than those who do not, and to pass on their behavioral type to their offspring. What will happen in such a population? If the cost of injury from losing a hawk-on-hawk fight is high in relation to the value of the resource, and some other assumptions are met, the population will reach a stable state where it contains a mixture of both strategies. Each type does well when it is rare. When hawks are rare they exploit the doves; when doves are rare they are the only ones avoiding damaging fights.

I said, "What will happen in such a population?" But the first thing to note is that natural populations are never as simple as this—there are no populations with exactly two behavioral types, where all the "hawks" are behaviorally equivalent to each other, and so on. Even theories that have a less obvious role for imagined scenarios often have some of this character; many evolutionary models assume populations that are effectively infinite, that deal with a uniform environment, and have unrealistically simple genetics. I will describe all models that make use of deliberate simplifications as *idealized*. Idealization can be contrasted with *abstraction*, which does not involve imagining things to be simpler than they are, but merely leaving some factors out of a description. Abstraction, to some degree, is inevitable; you can't include everything. Idealization, in contrast, is a choice. The border between these two is not always obvious, though, and will be important to some issues discussed later in this book.

It is hard to work out, philosophically, how to think about modeling of a kind that seems to involve the investigation of imaginary systems. One approach is to treat it as analogous to work that uses scale models: sometimes a model system is built, and sometimes it is just imagined. This takes us into problems about fictions and possibilities. A good model system is similar to its target; how can a target be similar to something that does not exist? A different approach is to see a "model system" as an abstract mathematical object. One way or another, any analysis of modeling has to grapple with the importance of consideration of the merely hypo-thetical or possible. As R. A. Fisher, who developed some of the most influential models of evolutionary change, put it many years ago (1930, p. ix), "The ordinary mathematical procedure in deal-ing with any actual problem is, after abstracting what are believed to be the essential elements of the problem, to consider it as one of a system of possibilities infinitely wider than the actual. . . ."

The approach I will take is to set aside some questions about what models are, and focus on the *products* of this work. The usual product of a piece of scientific model building is a set of *conditional* statements, statements of "if . . . then . . ." form. Con-ditionals raise philosophical problems of their own (Bennett 2003), but I am going to take them for granted. In modeling, the "if . . ." can be freely invented—modelers can explore any scenario they like. But the choice is usually guided by two goals. First, the scenario should be one where it is possible to work out, in some rigorous way, what would happen if it obtained. The obvious way to do this is to make the scenario one whose consequences can be investigated by mathematical analysis, or by programming a computer. The second goal, which can pull against the first, is that the scenario specified should be one that is usefully close to the real world.

In making the transition from "if" to "then," computer sim-ulation has become more and more important. People some-times describe model systems, such as evolving populations or predator-prey interactions, as being "inside" computer simula-tions. Rather than trying to make sense of that kind of claim, the way to understand simulations of this kind is to see computers as aids to the rigorous use of the scientific imagination. A computer is a physical device whose operation can be exploited to trace out very complex networks of "if . . . then . . ." relationships. A mod-eler will specify a setup, some relevant configuration of organisms or cells or something else, and then look for a way to determine the consequences of the setup. Computers are useful because our ability to specify these setups outruns our ability to work out how they would behave. Regularities in the operation of the computer can be used to tell us the consequences of the scenario that has been imagined. That this is the role of computers is illustrated by the way modelers move freely back and forth between "analytic" methods (solving equations) and simulations.

Whatever method is used, the typical result is a claim of the form, "*If* there are one-on-one contests over resources and these further conditions are met . . . , *then* the population will come to contain a stable mixture of hawk and dove strategies." Given that a modeler has to start by making deliberate simplifications, there are two ways to try to give conditionals as much relevance to the actual world as possible. One is to minimize the departure from reality on the "if" side, thereby retaining some of the problem of the real world's complexity. The other is to start further away, but also look for ways to then make the "if" side as logically *weak* as possible—that is, as undemanding or easy to satisfy as possible. A good way to do this is to develop many variants of a model, each of which makes different assumptions—all the variants are idealized, but in different ways. If things go well, many variants will lead to the same outcome. In the best-case scenario, the modeler starts out with assumptions that involve significant departures from reality, but is then able to make the "if" side of the conditional so undemanding that the actual world is one of the ones that satisfies it, or is very close to satisfying it.

An example is "Volterra's Principle," which says that in a system with a predator and prey population, if some external factor is introduced that kills them both, such as a pesticide, this will increase the relative abundance of the prey population (Wilson and Bossert 1971, Weisberg and Reisman 2008). Volterra started out making a lot of deliberate simplifications (Kingsland 1995), but many would say that the resulting principle (more carefully formulated than above) is a true generalization about actual systems, one that explains why the application of pesticides in agriculture often makes problems worse, as the pesticide does more harm to the natural enemies of a pest than to the pest itself. Some think that many conditionals in biology, especially ones like this, include a tacit clause requiring *ceteris paribus*, which means "with other things equal." The idea is that intrusions into the system from outside, and freak events, are set aside as irrelevant.

One attitude in this area holds that modeling always aims to bring us eventually to a description that does not idealize.

23

Ideally, there would be no idealization. Another view is that even when all the details can be known, idealized models are useful because they can highlight similarities between different systems. Richard Levins, yet another influential modeler, argued that science will always make use of models that simplify, and will retain several models of any given system, as a result of facts both about nature and about ourselves:

> The multiplicity of models is imposed by the contradictory demands of a complex, heterogeneous nature and a mind that can only cope with few variables at a time; by the contradictory desiderata of generality, realism, and precision; by the need to understand and also to control; even by the aesthetic standards which emphasize the stark simplicity and power of a general theorem as against the richness and the diversity of living nature. These conflicts are irreconcilable. (1966, p. 431)

Models that apply to particular cases with great precision are good, and so are models that cover a wide range of cases. Pursuing one of these goals usually requires sacrificing the other. Simplicity is good, too, and simple models can sometimes be applied to a wide range of real systems—but only if the model is interpreted as fitting these real systems in a loose or approximate way.[8]

Suppose it is agreed that conditionals are the typical results of modeling work. Perhaps *these* are the "laws of biology"? They are generalizations, not restricted in space and time, and when the connection between antecedent ("if . . .") and consequent ("then . . .") is established mathematically, they surely have a high degree of resilience—perhaps even necessity.

One possible disanalogy between these conditionals and laws has been discussed by Elliott Sober (1993, 1997). He thinks that biology does have laws, uncovered by modeling, but these laws are not empirical. They are just pieces of mathematics, and hence are necessarily true. Laws of nature are usually seen as having

[8]For detailed discussion of these trade-offs, see Matthewson and Weisberg (2009).

empirical content, but Sober thinks we should get used to the idea that laws can be purely mathematical. However, I do not agree that the conditional statements we get from models are mathematically necessary. Modelers might use mathematics to work out what follows from a set of assumptions, but the conditionals they end up with do not have purely mathematical content. Compare: "7 + 5 = 12" is mathematically necessary, but "if you put seven marbles on a table and add five there will be twelve marbles on the table" is not mathematically necessary. Whether this is true depends on the physical characteristics of marbles and tables. The same applies to conditionals about what will happen in an ecological system where a certain kind of predator eats a certain kind of prey. The mathematics is often where the hard work is done, but the conditionals that result are not merely mathematical statements. They are statements about the behavior of organisms, populations, and other biological objects.

Something that does look like an important difference between many of these conditionals and laws in a traditional sense is that the antecedent often describes a situation that does not actually occur. It is only close to something that occurs. Laws have traditionally been seen by philosophers as applying more directly to real systems—having antecedents that are often literally true—rather than merely making claims about what *would* happen in a nonactual scenario. In some cases, like Volterra's Principle, this might not be much of an issue, but many conditionals derived from models do have an idealized character. At this point, though, it is worth casting a more critical eye on the assumptions being made about laws in sciences like physics. Some think that idealization is so pervasive that theoretical "laws" in physics rarely or never describe the behaviors of actual objects (Cartwright 1983, Giere 1999). To the extent that this is true, the apparent contrast with biology fades.

Putting things together, there are two kinds of generalizations in biology that both look a bit like laws. First, there are conditional statements derived from models—these are not beholden to historical contingency, are often very abstract, and tend to idealize to some extent. Second, there are general statements about what actual organisms are like—spiders are carnivorous,

25

Mendel's First Law—that depend on historical contingencies and usually have exceptions.[9] Philosophers have often seen natural laws as independent of historical contingency *and* applying directly to real systems *and* highly general *and* having a kind of necessity. One possible position is that physics does have laws with this remarkable combination of properties, and biology does not. If so, perhaps the difference between these sciences is permanent, a consequence of the subject matter, or perhaps the gap will close. Another possibility is that laws in that sense are not found anywhere in science.

Setting laws aside, I will mention one other feature of model building before moving on. The aim of a modeler is often to come up with something whose departures from the real world do not matter too much: the "if . . ." is close enough to reality for the "then . . ." to be something we can expect to actually happen, at least approximately. Following the "internal logic" of a hypothetical scenario may become a goal in itself, however. This can lead to great theoretical creativity, but also to problems. After the financial crisis of 2008, the biggest crisis for banking and commerce since the Great Depression, some writers argued that economics had failed to predict and prevent the problem because it had become obsessed with the development of idealized models and had lost contact with reality. Paul Krugman, who had earlier won a Nobel Prize in economics, argued that the economics profession went astray "because economists, as a group, mistook beauty, clad in impressive-looking mathematics, for truth" (2009). Clarifying

---

[9] Within logic, generalizations about actual cases are also usually seen as conditionals: *if something is a spider, then it is carnivorous*. It is common then to distinguish several kinds of conditionals. A *material* conditional merely describes the layout of the actual world: *if something is a spider, then it is carnivorous* is true so long as there are no noncarnivorous spiders. This could be because there are no spiders at all. A *subjunctive* conditional asserts a connection between the two properties that goes beyond—somehow—the facts about which things actually exist. They can be expressed in a way that emphasizes this by saying, *if something were to be a spider, then it would be carnivorous*. There are also other kinds of conditionals, and the relations among them all are controversial: see Bennett (2003) and Edgington (2008). Here I will work within the idea that there are two (or more) kinds of scientific generalization, described in the text, without committing to an analysis of how they look from the point of view of logic.

this, the economists were probably making lots of true "if . . . then . . ." claims about markets and finance, using their high-powered mathematics, but the "ifs" were further from reality than they realized, and so were the "thens."

FURTHER READING

On laws, Armstrong (1985), Carroll (2004); in biology, Turchin (2001), Waters (1998), Ginzburg and Colyvan (2004), McShea and Brandon (2010); on mechanistic explanation, Craver (2009), Glennan (2002) and see note 4; on emergence, McLaughlin (1992), Bedau (1997), Bedau and Humphreys (2008); on idealization, Weisberg (2007a); on models, Wimsatt (2007), Frigg (2010), Downes (2011), Toon (2012), and see note 6.

Evolution and Natural Selection

A LARGE PROPORTION of the philosophy of biology is about evolutionary theory, as this part of biology unifies much of the rest, has a great deal to say about our place in the universe, and gives rise to many puzzles. Evolutionary change occurs at several scales. A standard way this is recognized is with a distinction between *microevolution* and *macroevolution*. Roughly, microevolution is change within a single species, and macroevolution is change in a collection of these units—a collection of species. This terminology makes the divide sound sharp, but rather than a situation where there are two distinct levels in nature, one can continuously "zoom in" and "zoom out" of what is going on in some region of space and time. As we do this, different patterns become visible. At a macroevolutionary scale, we find the "tree of life," a pattern of ancestry and descent linking all species on earth. Zooming in, we find change within the segments or twigs of the tree.

These relationships are represented in a diagram by the biologist Willi Hennig, reproduced in a modified form in Figure 3.1. Three scales are shown at once. At the most coarse-grained level, one species splits into two, giving rise to *phylogenetic* relationships between those species. Zooming in, this event is seen to be composed of many events involving relations between individual organisms, reproducing sexually. Change within each species is microevolutionary change. Zooming in still further, we encounter change within the life of a single organism. Those *ontogenetic* relationships are the subject of developmental biology.

3.1. EVOLUTION BY NATURAL SELECTION

In modern biology many concepts are important in explaining change within populations, but the one that generates most

Species differences

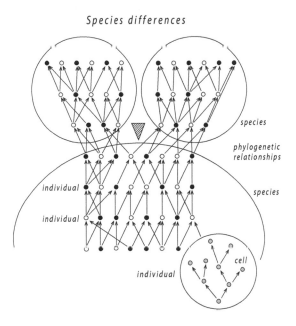

Figure 3.1. Change is represented at three scales. A species splits into two, a break in a "fabric" of individual organisms tied together by sexual reproduction. Differences in the reproductive success of individuals are seen within the fabric. In Hennig's diagram, change within the life of a single organism, at the lower right, was represented as a sequence of stages. I have replaced the stages with cells, linked by cell division. (Figure adapted from Willi Hennig's *Phylogenetic Systematics* ©1979 by the Board of Trustees of the University of Illinois. Used with permission of the University of Illinois Press.)

controversy is *natural selection*. One of Darwin's breakthroughs was to see that a huge amount can be explained in terms of the repeated action of a simple set of factors. Here is one of his summaries, followed by a passage from the end of *On the Origin of Species*:

Can it, then, be thought improbable . . . that . . . variations useful in some way to each being in the great and complex battle of life, should sometimes occur in the course of thousands of generations? If such do occur, can we doubt (remembering that many more individuals are born than can possibly survive) that individuals having any advantage, however slight, over others, would have the best chance of surviving and of procreating their kind? On the other hand, we may feel sure that any variation in the least degree injurious would be rigidly destroyed. This preservation of favorable variations and the rejection of injurious variations, I call Natural Selection. (1859, pp. 80–81)

Thus, from the war of nature, from famine and death, the most exalted object which we are capable of conceiving, namely, the production of the higher animals, directly follows. (1859, p. 490)

Compare this to a more recent summary by the geneticist Richard Lewontin:

A sufficient mechanism for evolution by natural selection is contained in three propositions:

1. There is variation in morphological, physiological, and behavioral traits among members of a species (the principle of variation).
2. The variation is in part heritable, so that individuals resemble their relations more than they resemble unrelated individuals and, in particular, offspring resemble their parents (the principle of heredity).
3. Different variants leave different numbers of offspring either in immediate or remote generations (the principle of differential fitness).

[A]ll three conditions are necessary as well as sufficient conditions for evolution by natural selection. . . . Any trait for which the three principles apply may be expected to evolve. (1985, p. 76)

The two summaries have different forms, as well as using different language. Darwin's summary makes generalizations about actual species. Lewontin's is best read as a conditional statement; *if* a species has the three features he lists, then evolution will occur. Another difference involves Lewontin's second condition. Parents, he says, need to resemble their offspring. They might not resemble them greatly, as long as they resemble them more than they resemble unrelated individuals. In Darwin's summary here (though not in all the summaries he gave) this seems to be taken for granted; favored individuals will "procreate their kind." But it is possible for a useful new trait to arise, help the organisms that bear it, and *not* be inherited, in which case there is no reason for the population to change.

Neither summary says that variation has to appear "randomly." Natural selection can work in a situation where new variations tend in some direction, perhaps even toward useful traits. But new variation *can* be produced in a random, haphazard, or "blind" way, and natural selection will sift the good from the bad. In Lewontin's summary there is no reference to a "battle for life," as in Darwin; whether or not there is a battle, change can occur if some do better than others. Neither summary says anything about genes or other mechanisms for inheritance. That is not surprising in the case of Darwin, but Lewontin, a geneticist, also treats genes as optional. Finally, both summaries make it clear that change is driven by local, short-term advantage, not by any kind of progressive tendency or foresight. Evolution occurs through the accumulation of routine events—births, lives, matings, deaths.[1]

Lewontin's summary gives three conditions and says they are necessary and sufficient for evolution by natural selection. Is it true that *whenever* you have these conditions, a population will change? Not in every case. Once you allow that the pattern of inheritance can be noisy, it is possible for the pattern of inheritance to push in one direction while the fitness differences push

---

[1] I treat *sexual selection*, in which some individuals have features enabling them to achieve more matings than others, as a kind of natural selection, not as something distinct.

31

in another, leading to no net change. For example, suppose the taller individuals have slightly *more* offspring than shorter ones, but taller individuals also tend to have slightly *shorter* offspring than themselves while short individuals do not. The two can cancel, leaving the population as it was.[2]

One response to this is to say that any conditional about change (except perhaps in basic physics) includes a *ceteris paribus* clause—a requirement of "other things being equal." Perhaps, but I think something else is being illustrated. There is a trade-off operating. If we make definite assumptions about the pattern of inheritance, it's possible to give definite statements about how differences in reproduction will lead to change. But any description like that will cover only some cases. If we want to say something that captures *all* cases, the summary won't have the same causal transparency.

In chapter 2, I distinguished two kinds of general claims in biology that look to some extent like "laws." There are general statements about actual cases, and conditionals that assert what *would* happen if a certain setup was realized, whether or not this ever happens. We see a distinction of that kind here.[3] Darwin's summary, though expressed using questions, is an attempt to describe facts about actual species. We could modernize it, like this: "In every species on earth, variation continually arises. Some of these new traits tend to be inherited across generations, and some inherited traits are beneficial to survival and reproduction while

---

[2]Here is an example that is about as simple as possible, modified from one by Robert Brandon. Suppose a population has four individuals, two large (L) and two small (S). They reproduce asexually. Two generations, with parent-offspring relations represented with arrows, are pictured here:

$$S \quad S \quad L \quad L$$
$$\downarrow \quad \nearrow \quad \downarrow \quad \downarrow$$
$$S \quad S \quad L \quad L$$

There is variation. Offspring tend, imperfectly, to resemble their parents. There are differences in reproductive success. But the new generation is the same as the old. Although there is heredity in Lewontin's sense, the pattern of inheritance itself pushes from large to small. This cancels the effect of the differences in reproductive success.

[3]Comments by Andreas Keller influenced my discussion in this paragraph.

others are not. In many cases, the traits beneficial to survival and reproduction become more common, while less useful traits are lost. This leads to ongoing change in the features of organisms in all species." It is also possible to look for a conditional: *If* such-and-such conditions hold, *then* a population will change, *guaranteed.* For example, if there is variation in a population, reproduction is asexual and offspring are exact copies of their mothers, everyone lives for the same length of time and reproduces at once, no one enters the population from outside or leaves, and individuals with some traits reproduce more than others, then the population will change. This is a verbal version of a mathematical model called the *replicator dynamics*, described in Box 3.1. This is sometimes seen as a foundational model of evolution (Nowak 2006a), and in a sense it is. But when applied to any real system, the model is an idealization, a deliberate simplification. Part of what Lewontin wanted to do in his summary is recognize that in many cases where the pattern of inheritance is noisy, evolution by natural selection can still occur. When you aim for generality of that kind, covering a wide range of systems, it is hard to make definite predictions. The replicator dynamics, on the other hand, is simple and gives precise predictions, but it is not a very realistic description of actual cases. The trade-offs operating here illustrate some general points made in the previous chapter (§2.3); descriptions that have the "simplicity and power of a general theorem," as Richard Levins put it, tend to be at odds with the "richness and the diversity of living nature."

Many debates about natural selection involve the concept of *fitness.* Evolution by natural selection is often said to be a matter of change due to fitness differences. The ordinary, nontechnical use of the term suggests two things, some sort of *fitted-ness* of an organism to its environment, and a kind of health or vigor. Talk of fitness was introduced to evolutionary theory in the 19th century by Herbert Spencer (1864), with the first of these meanings in mind. The term acquired a more technical role in the 20th century; or rather, it acquired several roles.

Lewontin's summary includes a "principle of differential fitness." But all Lewontin said was that some individuals "leave different numbers of offspring" than others. What if it is an *accident*

33

that some do better than others? Most evolutionary theorists recognize a distinction between change due to natural selection and change due to "drift"—accidental or random events that involve some individuals reproducing more than others. Lewontin seems to ignore this distinction. Here, in contrast, is a summary of natural selection by Alexander Rosenberg and D. M. Kaplan (2005, with their symbolism reduced a little here).

> *Principle of Natural Selection:* For all reproducing entities x and y, all environments E, and all generations n: if x is fitter than y in environment E at generation n, then probably there is some future generation n', after which x has more descendants than y.

Rosenberg and Kaplan treat fitness as something that *leads* to reproductive success. Note also that Rosenberg and Kaplan do not mention heritability, so they are focusing on just a part of what is covered by Lewontin.

The term "realized fitness" is often used for the actual reproductive output of an organism or a type of organism. The most influential way of understanding fitness in the other sense, the sense in which fitness explains or gives rise to reproductive success, is to see an organism's fitness as a *propensity* to have a certain number of offspring (Brandon 1978, Mills and Beatty 1979). A propensity is a tendency or disposition that can be described in terms of probabilities. A fair coin has a propensity to come up heads on roughly half the occasions it is tossed, even though it might always come up tails. Similarly, a fitter organism has a propensity to have more offspring than a less fit one. More technically, an organism's fitness can be seen as its *expected* number of offspring, where this expected value is calculated with probabilities that are interpreted as propensities. If an organism has a half chance of having no offspring and a half chance of having ten, its expected number of offspring is five. Very different organisms might have similar propensities to be reproductively successful.

Two kinds of problems arise with this view. First, there are cases where the expected number of offspring is not a good predictor of evolutionary change. I won't discuss those issues here.[4] The second is that propensities are rather strange features of the world. Any organism has a realized fitness, its actual number of offspring. It has zero, one, ten, or whatever. That outcome is the result of all the actual events in its life, all the causal details. Do we have to believe that behind that number there is some *other* number of offspring that it was "expected" to have, where that number is not merely a reflection of our ignorance of details, but a real feature of the world?

We might believe this, but it surely seems optional from the point of view of evolutionary biology. If someone thinks that realized fitness is the only kind of fitness that makes sense, this person does not have to stop believing in natural selection. The situation, as I see it, is like this. In the Lewontin summary, the Rosenberg/Kaplan summary, and others, the term "fitness" is applied to different parts of a causal sequence that biologists generally agree about. They agree that organisms live in different environments and have different ways of making a living. They agree that in all these cases, variations arise that in *some* way or other lead to an advantage in survival and reproduction. "Advantage" might be understood in terms of probability, or in some other way. In some cases where a new trait gives the organisms that bear it an advantage in survival and reproduction, those organisms will actually have more offspring. If the trait is heritable, then in many cases the population will change. All that is common ground. Talk of "fitness" is sometimes applied to the possession of a particular structural or behavioral feature that is useful in the case being investigated, sometimes to a propensity to succeed, and sometimes to actual reproductive success. A biologist might be wary of *all* talk about probability when dealing with macroscopic events, thinking that probabilities are just reflections of our ignorance. Indeed, I think it is reasonable to be a bit suspicious of standard distinctions between change due to natural selection and change due to "drift" or "accident." What we call "accidental" and "random" events have ordinary physical causes (unless we are talking about events at the microphysical level, which may be fundamentally indeterministic). Sometimes there is more regularity, more

[4]See Gillespie (1977), Sober (2001), Abrams (2009).

of a pattern, in who does well and who does badly, and sometimes there is less. Someone who is skeptical about standard distinctions between selection and drift might want to talk of fitness only in the "realized" sense, as seen in the Lewontin summary.

BOX 3.1. MODELS OF EVOLUTION
BY NATURAL SELECTION

The simplest mathematical model of evolution by natural selection is the "replicator dynamics" (Taylor and Jonker 1978, Weibull 1995, Nowak 2006a). Suppose there is a large population containing just two types, A and B, with frequencies p and $(1-p)$, respectively. Individuals reproduce asexually and simultaneously, with the parents dying right after reproduction. Then if W_A and W_B are the average numbers of offspring produced respectively by the A and B types, the new frequency of the A type after one generation, p', is related to the old frequency by this rule: $p' = pW_A/(pW_A + (1 - p)W_B)$. This model assumes that both types copy themselves exactly when they reproduce, and that other factors such as mutation and migration into the population are absent. It also treats generations as discrete steps. Other versions of the replicator dynamics treat time as continuous, not as a sequence of steps. The case with large and small individuals discussed in note 2 of this chapter where there were fitness differences and heredity but no change does not fit the assumptions of this model, as an L gave rise to a S. When applied to almost any real system, even asexual organisms like bacteria, this model is an idealization.

A more general way of representing evolution is with the "Price equation" (Price 1970, 1972, Okasha 2006, Frank 2012). This framework is more complicated than the model above, in part because it approaches populations in a different way, by tracking every individual and describing the statistical relations between "before and after" states. Assume there is an *ancestral* collection and a *descendant* collection of individuals, where all individuals can be described in terms of their value of a quantitative characteristic, Z (which might be size, for example), and assume a relation (usually interpreted as reproduction, though it can be understood in other ways, including persistence) linking the ancestors to their descendants. The aim is to represent the difference between the descendant and ancestral collections in their average values of Z, a difference represented as $\Delta \bar{Z}$. One version of the Price equation is this: $\Delta \bar{Z} = Cov(Z,W) + E_W(\Delta Z)$. Here $Cov(Z,W)$ is the covariance in the ancestral population between each individual's value of Z and their value of W, which is the number of descendants that individual is connected to, divided by the average number of descendants that ancestors have. So this first term on the right-hand side, sometimes called the *selection* term, represents the role of fitness differences; do individuals with a high value of Z have more (or fewer) descendants than others? The term $E_W(\Delta Z)$ measures the average change in Z that occurs between ancestors and the descendants they are connected to, where the average is weighted by the relative fitness of each ancestor. This term represents the role of the inheritance system.

This model does not assume copying, and the equation can be applied to sexual reproduction. It is an *abstract* description of evolution, leaving many things out, but not an *idealized* one; it can be applied to real cases without simplifying them. Unlike the replicator dynamics, though, the output of the equation cannot in every case be fed back into the equation as a new *input*, giving a model that applies over many time steps.

The example with fitness differences and heredity but no change in note 2 of this chapter can be described with a Price equation. Think of the large individuals as having the value $Z = 2$ and the small ones as $Z = 1$. The effect of the first term, which represents the effect of differential reproduction, is exactly balanced by the second term, which represents the failure of offspring to resemble their parents. So $\bar{Z}$, the average value of Z, remains unchanged. A Price equation can be used to represent evolutionary change at several part-whole levels in a system simultaneously, as the term on the far right-hand side can often be broken down into two terms that represent the roles of fitness differences and inheritance in entities at a lower level.

3.2. ORIGIN EXPLANATIONS AND DISTRIBUTION EXPLANATIONS

Natural selection is often described as the key to understanding how complex organisms can come to exist as a result of natural processes. But natural selection is also often described as a "filter": once variations have arisen, a few are kept while others are lost. A process of filtering cannot create anything, and assumes the existence of the things being filtered. Is it a mistake to think that selection can have something like a *creative* role in evolution?

The view that Darwin discovered a purely negative factor has been expressed often. An early example is Hugo de Vries, a biologist at the turn of the 20th century who was important in the history of genetics. De Vries noted that "in order to be selected, a change must first have been produced" (1909).

> [Natural selection] is only a sieve, and not a force of nature, not a direct cause of improvement. . . . [W]ith the single steps of evolution it has nothing to do. Only after the step has been taken, the sieve acts, eliminating the unfit. (1906, pp. 6–7)

To look more closely I will introduce some terminology, distinguishing between *origin explanations* and *distribution explanations*.[5] When we give a distribution explanation we *assume* the existence of a set of variants in a population, and explain why they have the distribution they do, or why their distribution has changed. Some variants may be common, while others are rare. Some may have been lost from the population, having been present before. An *origin* explanation, in contrast, is directed on the fact that a population has come to contain individuals of a particular kind *at all*. It does not matter how many there are; the point is just to tell us how there came to be some rather than none. So now we are explaining the original appearance of the variants that are taken for granted when giving a distribution explanation.

[5] This terminology is modified from one used by Karen Neander (1995).

Almost everyone agrees that natural selection can figure in distribution explanations. It initially seems that selection has no role in origin explanations, as selection can sort only things that already exist. This would not mean that evolutionary biology as a whole cannot give origin explanations. They would be given in terms of what we now call "mutation," along with the recombination of characteristics through sex. (De Vries was the person who introduced this modern use of the term "mutation.") Perhaps selection is a distribution-explainer while mutation is an origin-explainer.

I think, in contrast, that selection is essential to many origin explanations, and in a way that does give it a creative role in the evolutionary process. Part of Darwin's achievement was seeing this fact, and he was, as far as we know, the first person who saw it.

Selection is not an immediate, or proximate, cause of a new variant. The most important immediate sources of new variations, again, are mutation and recombination. However, natural selection can reshape a population in a way that makes a given variant *more likely to be produced* by the immediate sources of variation than it otherwise would be. As selection changes the background in which mutation and recombination operate, it changes what those factors can produce.

Suppose we are explaining the evolution of the human eye. Building the genetic basis of the human eye involved bringing together many genes. Consider a collection of genetic material, X, that has everything needed, as far as genes go, to make an eye, except for one mutation. So this background X is such that *if* a particular new mutation arises against X, it will finalize the evolution of the eye. Initially, X was rare in the population—it was the product of a mutational event that produced X from another precursor, W. Selection can make the appearance of the eye more *likely* by making X more *common*. This increases the number of "independent experiments" where a single mutation can give rise to the eye. If X remains rare in the population, then additional mutations are much less likely to produce the eye, as the right mutation has to occur in exactly the right place—in an lineage where X happens to be present. Selection makes the eye accessible to mutation in a way it would not otherwise be.

In that example I told the story working backward from a trait of interest. The process itself runs forward, without foresight, and involves many of these steps. When I call something an "intermediate" or "precursor," these terms apply only in retrospect, and the story can also be told without them. There is a population at time *t*, which contains variation. Some traits are useful to the organisms that bear them and others are not. They are useful for what they do at time *t*, not for what they might lead to later. The useful ones increase and their increase creates many sites at which further new variants arise. *Whatever* is favored at time *t* changes the background in which further mutations appear. Sometimes this process leads nowhere that strikes us as noteworthy, but sometimes it produces eyes and brains.

So selection can have a creative role, even though it is true in every case that "in order to be selected, a change must first have been produced," as de Vries put it.[6] The point can be made even more simply: if you can get to *Y* easily from *X*, but with difficulty from *W*, then you can make *Y* more likely to arise by having lots of *X* around and few *W*, as compared to the situation where you have lots of *W* and few *X*. As Patrick Forber (2005) notes, in a biological context this usually requires that trait *Y* be the product of many genes, or at least a lot of DNA. To the extent that a new trait can arise as a unit through a single change to any background, selection does not make it more likely to appear. But that is not how things are with eyes and brains, whose evolution involved changes to a great deal of DNA.

You might say at this point that it is not *selection itself* that does the originating; that is still due to mutation. Let's then make a three-way comparison, comparing mutation alone, selection alone, and mutation and selection together. Selection alone cannot produce new things, though it can keep the good ones that are already around. Mutation alone can produce new things, but in an indiscriminate way. There is almost no chance of it producing eyes and brains. Selection and mutation together can produce

---

[6] Compare de Vries to Herbert Spencer, in 1864: "To him [Darwin] we owe the discovery that natural selection is capable of *producing* fitness between organisms and their circumstances" (p. 446).

eyes and brains. So you might say it is *only the combination* of selection and mutation that is creative, and that would be fine. It might then be added that selection is *as* creative as mutation is. Perhaps that is exaggerating, though, as there is a tiny chance of mutation alone producing a complex new trait and selection alone cannot do that. And perhaps it is just wrong to say that the parts are "creative" when it is only the combination, selection plus mutation, that plays the crucial role. But the view that selection is only a distribution-explainer while mutation is an origin-explainer is wrong.

As discussed in the first chapter, Darwin had predecessors who glimpsed the idea of variation and selection but did not do much with it. One reason is that their hypotheses did not *iterate* the process. The cosmologies of Empedocles and Lucretius, for example, posited a special period at the beginning of the world in which variation appeared, followed by the culling of monsters. If there is no process where the results of selection feed back on *another* round of variation, there is no role for selection in explaining the origination of new structures.

A difference can also be described between Darwin's work and the "neo-Darwinism" that followed in the 20th century. Darwin's emphasis is on origin explanations. The distribution explanations he gives are simple: a new variant appears, and it either spreads or is lost. The iteration of many of these events explains how new kinds of organisms come to exist. From the 1930s onward, more sophisticated distribution explanations appear, made possible by Mendelian genetics. In writers like Fisher (1930), Haldane (1932), and Wright (1931), we see the idea of a discrete particle, a gene, inherited intact over many generations, coming into new combinations with other genes, and becoming more or less common—perhaps reaching a stable equilibrium frequency—in a gene pool.

I'll make a last point about origin explanations. Selection, I said, can make the evolution of eyes more likely by making eye precursors more common. But "common" is ambiguous—a trait might become more common in *relative* terms or in *absolute* terms. Natural selection is often described in terms of its effects on frequencies. Type *A* is favored by selection if it becomes more common relative to *B*, whether or not there are more *A*s

41

than there were before. That is fine in the context of distribution explanations, where one might either care or not about absolute numbers. But for selection to make the evolution of the eye more likely than it was before, it has to increase the *absolute* numbers of eye precursors. Then we see that the popular metaphor of selection as a "sieve" or "filter" is not a good one. In most cases of natural selection, some types decrease in numbers, and some types *increase*. Selection filters out some variants and amplifies others.

Suppose we have a population of eye precursors and non-precursors. Something that is not usually acknowledged here is the fact that the evolution of the eye could be made more common either by increasing the numbers of eye precursors alone, or by increasing the numbers of *all* types. In some environments, for a while, this can happen. When rabbits were first brought to Australia, in colonial times, they increased explosively in numbers.[7] The fittest rabbits certainly proliferated, but many of the less fit proliferated too. Eventually a situation is reached where if one type increases in numbers, another has to decrease. The "struggle for life," which Darwin emphasized, becomes relevant. In some modern discussions the struggle for life is treated as an inessential part of Darwin's theory, something that came from the influence of the pessimistic Malthus and Darwin's 19th-century context. But the fact of scarce resources—when it is a fact—ties relative reproductive success and absolute reproductive success together. Selection in Darwin's sense is as much an amplifier as a filter, and it is the amplifying that matters to its creative role.

3.3. Units of selection

The theory of evolution by natural selection, in the form discussed so far, is aimed at explaining how change takes place within populations of organisms. Organisms vary, organisms pass on traits, organisms differ in reproductive success, and the population or species changes as a result. But it was quickly seen

[7] There was a lag of about 70 years between their first introduction in 1788 and the explosive increase, which has apparently not been explained.

that a Darwinian pattern of explanation might be applied to other things. This is often explicit in summaries; Rosenberg and Kaplan's principle from section 3.1 is said to apply to all "reproducing systems." Here is another: "Evolution can occur whenever there are units of reproduction that produce other such units which inherit some characteristics of the parent units" (Doebeli and Ispolatov 2010, p. 676).

Some of the applications of this idea are inside biology: evolution by natural selection might operate at many levels in a hierarchy of parts and wholes; it might operate on genes, cells, organisms, groups, and perhaps species. Another set of applications lies outside biology. Natural selection has been seen working on ideas, technologies, economic firms in a market, and patterns of behavior in a culture. This section looks at the biological side (which is continued in chapter 5), and the next section looks at other applications.

Biology in the 20th century developed Darwinism by representing evolutionary change at the level of genes. This sometimes led to the idea that evolutionary change *is* change in frequency of genes in a gene pool. A rigorous version of this view was developed by George Williams (1966). Williams did so as part of a critique of explanations in terms of "the good of the species," and the good of other large units such as ecosystems and populations. Might cooperation, altruism, and restraint evolve because they make whole groups or species better adapted than their selfish rivals? No, said Williams; even if restraint or altruism does make a group "better" in some sense, that will not stop a selfish mutant from *invading* a harmonious group and flourishing at the expense of its well-behaved fellows. The lower-level process of competition within such a group will usually overwhelm any advantage the group might have as a unit.

As well as criticizing explanations in terms of group-level benefit, Williams argued that *all* evolutionary processes, even familiar ones in which organisms compete within a population, can be understood at the genetic level; in every case, one *allele* (alternative form of a gene) increases in frequency because it has some overall or net advantage over rival alleles at its *locus* (its place in the genome), as a consequence of the totality of effects the allele

43

has on cells and organisms that contain it. Richard Dawkins (1976) defended a colorful and grim version of this view, seeing organisms like ourselves as "gigantic lumbering robots" programmed by our genes (p. 21). For Dawkins, all evolution is the result of long struggles between selfish genes. Genes can persist, in the form of copies, while organisms and groups come and go "like clouds in the sky, or dust storms in the desert" (p. 34). Though organisms like ourselves are important parts of the living world, we are not *units of selection*, and whatever evolves is not for *our* evolutionary benefit, but for the benefit of our genes.

One reply to this argument is that although it is generally possible to "track" an evolutionary process by seeing what is happening to the frequencies of genes, it is not possible to *explain* what is happening by staying at the genetic level. Changes to gene frequencies are usually a result of the lives and deaths of whole organisms, and are sometimes affected by competition between larger units such as families and tribes. Most of the time, it is larger entities, like organisms, that natural selection "sees," not genes (Gould 1980, Sober and Lewontin 1982).

It is starting to look like there is an ambiguity of some kind, a failure to separate issues, behind the dispute. Modifying Dawkins's analysis, David Hull (1980) distinguished two senses of "unit of selection." In any evolutionary process, Hull said, two roles are seen. These roles may be occupied by the same things or by different things. First there must be *replicators*, things that are copied reliably over generations. Second there must be *interactors*, things whose activities and interactions with the environment affect which replicators are copied at a higher rate.[8] In the case of evolution in humans, genes are replicators and organisms are interactors. But sometimes groups or even species might be interactors, sometimes cells or genes might be. As for the replicators, these are usually genes, sometimes asexual organisms (for Hull, not Dawkins), and a few other things, but not organisms like us, because sexually reproducing organisms do not *copy* themselves. We pass on genes that are always entering into unique combinations.

[8] This is similar to Dawkins's concept of a *vehicle* (1982).

On this analysis, the people arguing that organisms and groups are important parts of evolutionary processes might be right about their importance as *interactors*, but this does not change the fact that genes are the *replicators*.

This view seems to clear up confusion and was embraced in both biology and philosophy (Lloyd 1988, Sober and Wilson 1998, Gould 2002). I used to think it is helpful, but now I think it is mistaken (though part of this framework will return with a possible new role in chapter 5). The quickest way to see there is something wrong is to look at the Lewontin summary given earlier. This summary had problems of detail, but it describes all that is needed for evolution by natural selection. And in that analysis, there are not two kinds of things, but one: the entities in the population that vary, inherit traits from their parents, and differ in reproductive success. If we have things with *those* properties, that is all that is needed. The passing on of "replicators" is one possible mechanism for inheritance, but it is optional. If genes are being passed from generation to generation, then genes might *themselves* satisfy the three conditions of variation, heritability, and differential reproduction, but to say that is to drop the replicator/interactor distinction and apply the same criteria to organisms, genes, groups, and everything else.

That I think is the right approach, but this is not the end of the units of selection problem; it will return in chapters 5, 6, and 8.

3.4. UNIVERSAL DARWINISM

Once the idea of variation and selection snaps into focus, it is tempting to apply it to many systems. The Darwinian dynamic, or something like it, has been seen in scientific change, technological change, individual learning, and elsewhere. One of the quotes I gave earlier summarizing evolution by natural selection— "Evolution can occur whenever there are units of reproduction that produce other such units which inherit some characteristics of the parent units"—is from an article about *religion*; it gives a theory of how religions compete and spread, and not through any kind of "religion gene" but by cultural processes. Are these

45

analogies superficial, perhaps even mistaken, or do Darwinian ideas help us to understand change in these other systems too?

It is possible to use Darwinism as the basis for a general theory of all change of a certain kind. I'll call this category *adaptive* change—change that involves improvement to the design of a system or its ability to deal with its environment. (The idea of adaptation will be treated more warily in the next chapter.) This broadening of Darwinism can be both explained and motivated by looking at examples. First, here is a theory of learning, in humans and animals, developed especially by Edward Thorndike (1911) and B. F. Skinner (1974): Organisms often produce new behaviors in a haphazard way, trying out new things from time to time. If a behavior has good consequences in a given situation, it is retained. When that situation arises again, the organism is more likely to produce the same behavior.

Second, here is an account of scientific change itself. Karl Popper (1959) thought that change in science occurs by an endless cycle of *conjecture and refutation*. Scientists imaginatively propose new theories, going far beyond the data, and then try to refute these theories by collecting further observational evidence. Conjecture and refutation; trial and error; mutation and selection.

A third example takes us back to biology. How do our bodies learn to recognize invading viruses and bacteria? How does "adaptive immunity" work in organisms like us? Early proposals had it that the body somehow receives the impression of the invader, like the stamping of a form on a wax tablet. Nils Jerne (1955) and Macfarlane Burnet (1958) proposed instead that the immune system uses a mechanism of variation and selection. It produces many different antibodies in a "random" way, and when a cell happens to make antibodies that can bind to an invader, these cells are made to proliferate at the expense of others. Later work on the immune response has complicated this picture, but the basic insight has stood up. The development of the brain in infancy and early childhood is also thought to work by a selection process, though here I think the analogy is becoming weaker: the way to wire up a growing brain is to start with *too many* connections between neurons, and trim away the ones that do not serve a useful role while strengthening the ones that do

(Changeux 1985, Edelman 1987). A more speculative theory in this family is the idea that cultural change, especially improvement in skills and technology, occurs by a process in which new ideas and behaviors pop up from time to time, and some spread because they are *imitated* more than others. I'll look at this one in chapter 8.

Given all this, it is tempting to offer a grand theory: *whenever adaptive change occurs, some process of variation and selection is responsible.* This is a kind of "universal Darwinism." Views along these lines have been developed by the psychologist Donald Campbell (1960), Daniel Dennett, in philosophy (1974), and others.

At one stage in the 20th century it must indeed have seemed that *everything* was turning out to be variation and selection. The argument is harder to make now. The Thorndike-Skinner theory of learning and Popper's theory of science are not widely accepted in their respective fields. Both are oversimplified, but that is not the heart of their problems. They overstate the importance of pure trial and error. Sometimes variation and selection at one time scale builds another system that can adapt to the world in ways that are *not* made up of more variation and selection (Amundson 1989). Evolution by natural selection built our brains, and maybe nothing else could. But once it has done so, our brains can do things that are smarter than just throwing out new behaviors—or beliefs— and seeing if they work. We can engage in logical reasoning and planning (at least some of the time), and shape ideas and behaviors without *exposing* them at every step. Sometimes variation and selection builds more variation and selection, as in the vertebrate immune system, and sometimes it builds something else.

So there is also a more moderate "universal Darwinism": whenever we have a system that can undergo adaptive change, there must be variation and selection *somewhere* in the story, but one variation-and-selection process can build machinery that creates further improvements by working differently. (Richard Dawkins, who coined the term "universal Darwinism," had something like this second view in mind.)

Setting aside the most ambitious views, it is fruitful to keep looking at the relationships between different selection processes. We can start by recognizing a very general category: systems in which there is variation, and where successful variants become more common or are more likely to be retained. A two-way distinction can then be made within this class. On one side we have cases where the way successful variants are retained is through reproduction *by* those entities. This includes biological evolution. On the other side are cases where the retention of successful variants is done by a more centralized mechanism. Trial and error learning is like this; a successful behavior does not *make more behaviors*. Rather, something in the brain registers the good results that came from a behavior and generates similar behavior on later occasions. To the extent that cultural change in a human society involves the retention and passing on of successful innovations, perhaps a mixture of both modes is present, along with other things. In the case of cultural change and elsewhere, though people often argue about whether such and such a process is or is not "Darwinian," what we find is many differences of degree. Processes can be more or less Darwinian, and can shift, either rapidly or slowly, with respect to this status. Several dimensions are relevant here. Is new variation produced in an undirected way? Does the way that variants are retained allow for the accumulation of small improvements? Do successful variants spread by reproducing, or in some other manner?

Once the power of an iterated variation-and-selection mechanism is seen, it is surprising that the history of theories of this kind goes *from* Darwin's biology *to* other applications. Evolutionary change in biology is slow and inaccessible, and the role of variation and selection there must be detected among a great deal of noise. Darwin was helped by analogies to plant and animal breeding in farming, but in learning by trial and error, or the spread of an invention by imitation, everything happens on a time scale that is much easier to observe. Once attention is drawn to these things, there can be no doubt of the role of *some* kind of variation and selection. But—to pick one lineage as an example—the theories of learning and knowledge in people like Locke, Hume, and Kant missed this idea entirely.[9] There is an alternative

---

[9] Here is a point of near-contact in David Hume, writing about social patterns: "Two men who pull the oars of a boat, do it by an agreement or convention,

history, another possible world, in which Darwin or someone was able to draw on an understanding of variation and selection in a range of more obvious areas and apply it to the less obvious case of biological change. In the actual history, the more difficult application came first, and others came later.

FURTHER READING

For relations between Darwin and modern views, Lewens (2010), Sober (2011); for fitness and drift, Ariew and Lewontin (2004), Walsh et al. (2002), Millstein (2006), McShea and Brandon (2010); for evolutionary explanation, Beatty (2006); for units of selection, Lloyd (2001), Okasha (2006); for general theories of selection, Hull et al. (2001); for other applications of Darwinism, Dennett (1995), Wilson (2002), Hodgson and Knudsen (2010).

---

although they have never given promises to each other. Nor is the rule concerning the stability of possessions the less derived from human conventions, that it arises gradually, and acquires force by a slow progression, *and by our repeated experience of the inconveniences of transgressing it*. . . . In like manner are languages gradually established by human conventions without any promise. In like manner do gold and silver become the common measures of exchange" (1739, bk. 3, pt. 2, §2, italics added).

Adaptation, Construction, Function

SOME SCIENTIFIC THEORIES bear directly on the place of humans in the universe's total network of causes and effects. Clear examples include materialist and determinist views, and their rivals. Other theories do it in less obvious ways. They might bear on whether we are here for a reason or by accident, or on whether our lives have a purpose over and above what we decide it to be. More subtly still, they can bear on whether our activities (perhaps as humans, perhaps as living things) are fundamentally *reactive*, responding to patterns and demands that originate outside us, or whether we, in some sense, call the shots, imposing structure on experience, and perhaps on the universe. This chapter is about a family of concepts with two roles in the philosophy of biology; they raise specific problems within biological work and also bear on larger questions about our place in the world.

4.1. ADAPTATION

Parts of Darwin's thinking were motivated by the need to explain *apparent design* and *adaptation* in the living world. These phenomena had been used as evidence for God as creator, especially by writers in the "Natural Theology" tradition (Paley 1802/2006), and any view that removes God from biology has to confront them. Several biologists have recently echoed Darwin on this point, saying that "the cardinal problem of biology" is to explain adaptation and apparent design.[1]

What exactly has to be explained? Saying that the problem is "apparent design" does not help much; what are the *real* features

[1] Gardner and Welch (2011); see also Dawkins (1986).

of organisms that make them *seem* designed? Some answers given to this question are rather metaphorical. The geneticist Dobzhansky used musical analogies. He spoke of adaptedness in terms of "harmony," and said that organisms are "attuned to the conditions of their existence" (1955, pp. 11, 12). Reaching past the metaphors, there seem to be two elements here. One is the *organized complexity* of living things, their containing parts that work together in a fine-tuned way (music again . . .). The other side of the issue concerns relationships between organisms and their environments; there seems to be a well-suitedness, a fittedness, of organisms to their circumstances of life.

These two raise somewhat different questions. First, how is organized complexity distinguished from *dis*organized complexity? The general discussion of organization in chapter 2 can be applied here, but what matters most is something more specific: in some systems the parts work together in a way that maintains the distinctive features of living activity. I'll say more about "living activity" in the next chapter. For now, what I want to emphasize is that this first side of the problem reduces to the question of whether biology can explain the origin of living systems and their elaboration into complex forms. That is a question that all evolutionary biology contends with, not a specific challenge that has to do with "design."

Let's next look at adaptedness. Here the path taken by much of the literature has been a bit surprising. Adaptedness seems to be a feature of organisms that leads to evolutionary success: better adapted organisms are more likely to survive and reproduce. To make a generalization like that, though, suggests that there is some single abstract feature, adaptedness, which is seen sometimes in ants, sometimes in fir trees, and sometimes in viruses, and which in all cases makes them likely to succeed. Once the question is raised, many people respond by thinking there can't really be such a feature. Talk about adaptedness and adaptation is fine, but these terms should be understood in a more minimal way. A particular trait is an *adaptation* if it has in fact been favored by natural selection, whether it seems to involve harmony with the environment or not (Brandon 1990, Sober 1993). An organism displays *adaptedness*, or a trait is *adaptive*, if it is *likely* to succeed under natural selection. In any particular case, there will be some reason why the organism is likely to reproduce or the trait is likely to spread, but these reasons will be different in each instance.

If someone then says that adaptedness is a general feature of the living world, what are they saying? They are saying that lots of organisms have particular features that make them likely to succeed, given their environments, in processes of natural selection. That is not saying very much, especially given that selection is a comparative matter. And if you want to explain the fact that trait *A* spread and trait *B* was lost, it is saying very little—not nothing, but not much—to say that *A* spread because it was better adapted or made organisms better adapted. It is saying that *A* succeeded because it had features that made it likely to succeed. The idea that high degrees of adaptedness are an observable phenomenon that need a special kind of explanation has disappeared. The same issue, nearly or exactly, arises with *fitness*, as discussed in the previous chapter, and some people would not distinguish the two. Whether we have one problem here or more than one, the situation is that there initially seem to be properties that involve an abstract "match" between organism and environment, that tend to lead to success. But when people try to say what these properties are, they end up deflating them.

It has sometimes been argued this is a problem for evolutionary theory as a whole, because the theory is committed to a principle—*the survival of fittest*, which I take to mean something like *the fittest organisms will survive*—that must be understood in a way that is both true and nontrivial. But no problem arises here for evolutionary theory itself. The following combination of views is entirely fine: given any organism's particular circumstances and lifestyle, there are traits it can have that will give it a local advantage, but there is almost no limit to what might be useful for some organism somewhere. Across space and time, different populations change as a result of the local advantages that some organisms have over others, but there is no general harmony-like feature that all the successful organisms share; each just has *some* features that work well locally. That is one Darwinian package of views. It is also possible to look for more than this, to look for abstract similarities across successful organisms of different kinds, from paramecium to redwood, and try to give a more general theory of what adaptedness involves. This might include efficiency in the use of energy, perhaps.[2] It might turn out that what makes for success is similar within broad classes of organisms, but different across them. This is an internal question for evolutionary theory, not something that threatens the foundations.

These questions about minimal and richer concepts of adaptation affect debates within biology about whether natural selection is, in some sense, the most important factor in evolution, deserving a central place in biological theory. Views like this are forms of *adaptationism* (Gould and Lewontin 1979, Godfrey-Smith 2001b). One way to hold such a view is to say that natural selection is ubiquitous and shapes everything around us. That can be called *empirical adaptationism*. A different view, harder to assess empirically, is that selection is the most important factor not because it is everywhere and all-powerful, but because it is present *enough* to enable evolutionary theory to solve a problem that could not be solved otherwise, the problem of how adaptation and apparent design can arise without God. This view, which can be called *explanatory adaptationism*, requires that adaptation and apparent design must be real features of the living world, not just features of the way that some plants and animals strike us. There is also a purely *methodological* form of adaptationism: perhaps the best way to investigate living things is to look first for adaptedness, as a way of organizing our work, even if we expect to often find departures from it.

---

[2]There is an intriguing relationship here between the problem of adaptedness and the problem of explaining *truth*, as a property of representations (discussed further in Godfrey-Smith 1996). Truth, like adaptedness, appears to be a sort of abstract "match" between many different representations and the world, which can be used to explain success. People with true beliefs tend to do better than people with false ones, even when their interests and their projects are very different. But much recent work on truth in philosophy has tended to deflate the concept of truth (e.g., Horwich 1990), either generally or with respect to its link to success, making the familiar idea that truth is desirable and can be a "fuel for success" rather mysterious.

A third controversy about adaptation involves the overall pattern of causal interaction between organisms and their environments. As I said at the start of this chapter, a number of theories in different parts of science and philosophy seem to offer a *reactive* view of organisms, or minds, or biological systems in general. They give theories of how we (and things like us) adapt to patterns and events that are generated outside and independently of us. These contrast with views that see us (and things like us) as *imposing* ourselves on our environments. Theories of knowledge, for example, differ on whether the role of the mind is primarily to respond to patterns in experience, or to impose structure on experience and perhaps on reality itself. This is one aspect of debates over "empiricist" views of mind and knowledge in the tradition of John Locke and David Hume, which emphasize the mind's responsiveness, and their clash with one kind of "idealist" view, seen in Immanuel Kant and G. W. Hegel (though these terms, "empiricist" and "idealist," are used in other ways as well).[3] Analogous debates have arisen in and around biology, and both philosophers and biologists have had their eye on these issues since the early days of evolutionary thinking (Pearce forthcoming). A detailed argument on this theme has been developed more recently by the geneticist Richard Lewontin (1983, 1991).

Lewontin argues that mainstream evolutionary biology has developed a picture in which organisms are passive in evolution, responding to environments that change independently of them. Environments impose demands on organisms, and organisms must adapt. Earlier in this section I said that people have often used metaphors to make sense of adaptedness as a relation between organism and environment. Another common metaphor illustrates what Lewontin has in mind; people say that an organism is adapted when it has good *solutions* to the *problems* posed by its environment. Lewontin regards this asymmetry—between environments that call the shots and organisms that respond—as

---

[3] Another quote from William James, expressing (not endorsing) this first general attitude: "Man is no law-giver to nature, he is an absorber. She it is who stands firm; he it is who must accommodate himself. Let him record truth, inhuman tho it be, and submit to it!" (1907, p. 16).

integral to the style of thinking in biology that emphasizes adaptation. This, he thinks, is a mistake. Instead of adapting to their environments, "organisms construct every aspect of their environment themselves" (1983, p. 104). Organisms are thus the active determinants of their own evolution.

Lewontin has expressed both what he sees as the error and the way to correct it with equations that represent change over time. The standard picture, he says, is one in which the state of an evolving population at time $t+1$ is a function of the prior state of the organisms—where they start—and the pressures imposed by the environment at time t. The environment will also change over this period, but in a way that is largely autonomous. Thus,

$$O_{t+1} = f(O_t, E_t)$$
$$E_{t+1} = g(E_t)$$

Here f and g are two functions, in the mathematical sense, and O is not the state of a single organism, but of a population or species. E is the environment. A better picture, Lewontin says, is one where not only is the state of the organisms at time $t+1$ a function of the environment at t, but the state of the *environment* at $t+1$ is a function of what the *organisms* are like at t.[4]

$$O_{t+1} = f(O_t, E_t)$$
$$E_{t+1} = g(O_t, E_t)$$

4.2. CONSTRUCTION

In the way of thinking advocated by Lewontin, construction of environments is an alternative to adapting to them. It is also possible to use these terms differently, where construction is not an alternative to adaptation but a variety of it. I will work first within

[4] I have replaced Lewontin's differential equations with difference equations. See Godfrey-Smith (2001a) for a more detailed discussion of Lewontin's arguments.

Lewontin's way of setting things up, and return to this at the end of the section.

Why should we think that organisms construct their environments? Lewontin draws on three kinds of phenomena:

1. Organisms *select* their habitats, actively choosing where to live.

2. Once they are in some particular place, an organism determines what is *relevant* to it. An organism's size and constitution determine whether, for example, a two degree change in temperature is something it must deal with or can simply ignore. In evolution, a lot depends on the statistical structure of the environment that an organism faces, the pattern of variation by which states come and go. But that statistical pattern depends on which distinctions between states are relevant to the organism's life.

3. Organisms physically *transform* their surroundings, by consuming resources, depositing waste, and rearranging objects around them. The oxygen-rich atmosphere around the earth, for example, might appear to be a "given," but it is the product of the metabolic activity of photosynthetic organisms.

All these phenomena are real, but what is their significance? The third is something that some organisms do more than others—beavers, worms, and coral-building invertebrates have enormous effects on their physical surroundings, while many birds and small mammals have less effect. The second phenomenon, in contrast, is not a matter of degree and is something no organism can fail to do. Also, it is more a matter of how an organism determines how the environment affects the organism *itself*, than how the organism affects its surroundings. Using the equations above, it has to do with the importance of O at one time in affecting O at later times, not the importance of O in affecting E. What about the first phenomenon, the choice of habitat by organisms? In a way it is between the other two. An organism by its behavior determines that some region of space counts as its environment, but it may do this without making changes *to* that part of the physical world.

For the rest of this section I'll look more closely at the physical transformation of environments by organisms. The terms "niche construction" and "ecosystem engineering" are sometimes used for this phenomenon (Odling-Smee et al. 2003, Jones et al. 1994). Some biologists argue that mainstream biology has persistently downplayed and underestimated this factor in evolution. One possible response to this is to deny that it has been downplayed at all. Game theory models (§2.3) and a range of other theoretical ideas emphasize the way that organisms' behaviors change the social environment in which behavior itself evolves. This is a special case of "environment," though. Another possible reply is that while no one would *deny* that organisms change their surroundings, for some purposes this fact can be put in the background. Evolutionary biology has spent more time describing the ways organisms change in response to their environments than vice versa, because the main phenomena in this area are not much affected by the causal arrow from organisms to environments. You can't describe everything at once.

A way to respond to this last claim would be to show that it matters to evolutionary explanation itself that organisms transform their environments. One argument that has been made is that the effects of organisms on their surroundings form a kind of inheritance system; the next generation inherits an environment shaped by its parents, as well as inheriting their genes. Another argument connects to ideas in the previous chapter.[5]

Origin explanations in evolutionary biology explain how new traits and new kinds of organisms come to exist. Natural selection, though it can appear to be purely "negative," can be important in the production of novel traits, because it changes the background against which mutation operates. That argument, made in section 3.2, assumed the existence of selection pressures; it assumed an initial state for some organisms, an environment they had to deal with, and worked forward from there. But where did the environment assumed to be present at time t come from? In many cases it was affected by the organisms at time $t-1$. The organisms at $t-1$

[5] For the first argument here, see Griffiths and Gray (1994) and Odling-Smee et al. (2003). Regarding the second, I am indebted to Adam See and Fiona Cowie.

affect the environment at t; the environment at t plus the state of the organisms determine the selection pressures; the population changes; and those changes affect what mutation can produce later on. So the actions of organisms on their environment at one time affect what mutation can give rise to later.

There is a connection between these themes and another way that organisms have been seen as "active" in evolution. Mary Jane West-Eberhard (2003) argues that mainstream evolutionary biology does not give a good account of how innovation in evolution occurs. She thinks that a neglected causal sequence is one in which organisms respond to a novel stimulus by *doing* something new, making use of their capacities for flexible behavior, and this alters the selection pressures that are relevant in that population from then on, leading to genetic change. "For these reasons I consider genes followers, not leaders, in adaptive evolution" (2005, p. 6547). This process does not require the transformation of the external environment, so it is distinct from the line of argument above. But it is a similar attempt to reorient our thinking about where causal chains begin, and about what is "active" in biological processes.

Here I have been talking in abstract terms, but there are particular species where these effects assume special importance. This includes ourselves; a huge amount of the environment confronted by any human now is the product of the activities of humans at earlier times. This includes both the social setting in which we live—"environment" in a special sense, as each of us is environment to the other—and the collection of enduring artifacts and reshaped parts of the physical world around us.

I'll put some of the ideas arising over the past few pages together. When claims about activity and passivity, following and leading, are made, it is often hard to work out whether there is anything substantive at stake, or the choice is merely over forms of description and accompanying pictures. I tend to think that often there are real issues at stake, including in this case. A resolution of any actual confusion can be achieved by first setting aside the problem terms ("adaptation," "construction") and describing the phenomena in other ways: organisms are embedded in environments and have to deal with them; the value of nearly any trait of an organism depends on what the environment is like and what the rest of the population is like. Some environmental factors place constraints on organisms, and some create opportunities. In response, organisms may respond by changing only themselves (over their lifetime or over evolutionary time), or by also transforming their environment. If an organism does not make physical changes to its surroundings, it may change the way it relates to that environment, doing this in such a way that a particular external feature no longer poses a problem.

I said at the start of the section that it was possible to see transformation of the environment as a *kind* of adaptation, or as an *alternative* to adaptation, and I set that aside and worked within the second terminology. What I've just said casts light on why the issue arose. The need to adapt behavior to circumstances, the environmental contingency of effective action, is a general fact. To say that is to use the term "adapt" in a broad sense. Transforming the environment is one way of dealing with this need to adapt to circumstances. There is also the possibility of "adapting" in a narrower sense, in which to adapt is to change oneself but leave external conditions unchanged. In this sense, adaptation is not so ubiquitous.

This discussion of the biological side of things may gesture to a resolution of some of the larger philosophical issues concerning "passive receptivity" and "active construction" as relations between self and world. Views tend to gravitate toward models of the pushing around of our minds by something "given," or to models of the self-propelled imposition by agents of structure on their world. What is real, though, is the environmental contingency of effective thought and action, the need to adjust action to circumstances, and also the fact that many effective responses to environments transform them.

4.3. FUNCTION AND TELEOLOGY

The family of "teleological" concepts includes the concepts of *goal*, *purpose*, and *function*. The function of something is what it is *for*. A goal is an outcome that something's behaviors are *aimed* at or *directed* toward. Many activities of living organisms seem to be directed toward goals, and many of their parts seem to be for something. What is the relationship between modern biology and this way of thinking about living things?

I'll start by sketching several possible pictures of the relationship. One is that there has been a simple replacement of one framework with another. Teleological concepts were central to the worldview of Aristotle and of many who followed him (§1.2). This teleological outlook on nature was gradually replaced by a more mechanical one, based on physical causation.[6] Immanuel Kant wrote at the end of the 18th century that it was absurd to hope for a Newton in biology, someone who would make the "genesis of a blade of grass" comprehensible without drawing on *intention* (1790/1987, §75). Not long after, Darwin did what Kant said would never be done. Or perhaps Darwin began it and 20th-century biology completed the task. On this first view, intelligent agents do have goals and intentions, but these agents are physical systems and their intentions are inner states that guide behavior. Darwinism explains why it *seems* that other living structures pursue goals, when in fact only blind mechanical forces are operating. Evolution is purposeless, organisms and their parts are not *for* anything, and teleology is an illusion except in cases where an intelligent agent is making choices for reasons.

A second view is that evolutionary theory showed that although God or some other supernatural factor is not the *source* of the functions and goals of biological structures, these properties nonetheless can be real. When some trait or structure has been selected for and maintained because it has effect X, it has X as its function. This applies to the traits of plants and bacteria as well as intelligent organisms. Modern biology has replaced one source of teleological properties with another, and has also reduced the number of things that can be understood in teleological ways, but this includes all products of evolution, not just intelligent agents.

---

[6] In chapter 2, I used a narrow sense of the term "mechanism," restricted to systems that are *organized*. In this section I use terms like "mechanical" in a broader, more historically standard sense, in which very unorganized systems can still be understood "mechanistically" if everything that happens in them occurs through local physical causation.

A third view, a middle road, is that teleological thinking is part of a "stance" that we apply, a way of looking at things and explaining them. It is often a useful framework, especially with evolved systems. We pretend that some physical objects were designed by an agent with intentions, and that the parts of these systems have functions and the activities of the system have goals, even though we know there was no real designer. We say these things because they are often useful.[7]

These first three views do not differ about how the world actually works. They agree that earlier teleological views about nature have been replaced. They differ about the exact commitments of the pre-Darwinian view, and hence about how much disruption there has been.

Views that contrast more sharply with these hold that teleological explanations are valuable, perhaps indispensable, while being in no way reducible or explainable in terms of ordinary physical causation. This defense of teleology, in turn, is occasionally made by arguing that there are deficiencies in our scientific picture itself, that physics must be augmented (Nagel 2012), but in other cases by arguing that it is something about *us*, our ways of understanding, that makes it inevitable that we describe living things in teleological terms, regardless of how this fits with what we learn in other parts of science (Thompson 2004). The aspect of this last view that is both most distinctive and hardest to defend is the claim of inevitability, the idea that these habits of thought could change only as part of a complete breakdown in our conception of ourselves as agents. The idea that teleological description is a "free-standing" approach to explanation also becomes less plausible if it is possible to say something in other terms about when and why teleological forms of description come to be useful, what sort of relation they have with the picture we get from other sciences.

With all this in mind, in the rest of this section I'll look at the place of teleological and *quasi*-teleological concepts within an evolutionary worldview. I'll focus on the concept of a function. A

---

[7]For the first view, see Ghiselin (1969, 1997); for the second, Wright (1976), Neander (1991), Godfrey-Smith (1994); for the third, Dennett (1987, 1995).

starting place is the idea that an object's function is what it is *for*. Modern biology certainly recognizes relatives of this idea. First and most minimally, one can talk about the typical causal role that something has, especially its typical contribution to a larger and more complex system. This is a low-key way to talk about functions, certainly unproblematic. Furthermore, a biologist might mark out and make assumptions about the *normal* operation of a system, while not doing this in a theoretically loaded way. A system might exhibit a set of behaviors that someone particularly wants to understand, and the functions of the parts will then be their contributions to those behaviors (Cummins 1975, Craver 2001). This is sometimes called the *systemic* concept of function.

A richer concept is also available. A part of a system, such as an organism, can have an effect that explains *why the part is there*. For example, pumping blood around the body is the thing our hearts do that explains why they are there.

Larry Wright (1973) suggested that this phenomenon is the key to making sense of the idea of a biological function; a function is an effect something has that "explains why it is there." This is an *etiological* concept of function. The explanations relevant here can involve various different selection processes, including those seen in biological evolution, imitation, and deliberate choice. Biological evolution is one of a family of processes in which there is some means by which the effects of something's presence and capacities feed back and affect its chances of being kept around, or affect the chances of similar things coming to exist in the future. Wright also emphasized the analogy between these cases and ones where it is not *actual* effects but *envisaged* ones that matter. For example, I can put a newly invented device into a car's engine to pump the fuel. Pumping fuel is then "the thing it does that explains why it is there," even before it has ever been turned on, so long as it is disposed to pump fuel and that is why I put it there. For Wright, both this case, and cases where some part of an organism has been maintained by selection because it *actually* works as a pump, are cases where a *consequence etiology* is present.

This language chosen by Wright seems to deliberately emphasize the initially mysterious feature of teleological explanation, the idea that the future can reach back and affect the past—an effect of something explains why it is around to have that effect. There is no mystery, though, if we separately consider how each kind of selection process works. In biological evolution, present-day hearts are around because of the effects that earlier hearts had; the ability of a present-day heart to pump blood had no role in its coming to exist, but the useful contributions to living activity made by the pumping of earlier hearts did. This is a kind of *feedback* process, operating over a long time scale. There are also other ways in which an object's effects lead either to its own maintenance or to the production of further objects of the same kind. This applies not only to the production of "objects," but also to a variable in a system taking on a particular value. Feedback processes of this kind were poorly understood in any context before the 19th century. This connects to the discussion at the end of chapter 3 of the timing of Darwinism in relation to other events in the history of ideas. Feedback and the 19th century also bring to mind industrial technology and the first mechanized control devices, including James Watt's "steam governor." Darwin himself did not, apparently, make any connections here, but Alfred Russell Wallace did, in the 1858 paper he sent to Darwin, prompting Darwin to finally publish.[8]

[A]ll varieties in which an unbalanced deficiency occurred could not long continue their existence. The action of this principle is exactly like that of the centrifugal governor of the steam engine, which checks and corrects any irregularities almost before they become evident; and in like manner no unbalanced deficiency in the animal kingdom can ever reach any conspicuous magnitude, because it would make itself felt at the very first step, by rendering existence difficult and extinction almost sure soon to follow. (1858, p. 62)

[8] This way of looking at selection also suggests a connection to the "invisible hand" of Adam Smith's economic theory (1776). This, too, seems to be a connection that Darwin did not notice, or at least did not note. Darwin did have knowledge of Smith's work; see Schweber (1977) for connections between Darwin and the Scottish school that included Smith.

The puzzling impression given by teleological explanations that the future can affect the present is removed by noting the way that natural selection and other kinds of feedback enable present effects to contribute to the maintenance of a state of a system, the continued existence of an object, or the production of more objects or states of the same kind.

The alignment between older teleological concepts and the properties of things that can be understood in terms of selection and other feedback processes is imperfect. To some extent the match can be improved by adding detail. For example, the term "function" is usually applied only to effects that contribute to the success or well-being of a larger system. It is possible for something to have an effect that "explains why it is there" without any such helpful contribution to a larger system—consider a parasite disguising itself within a host's body. In response, one might say that a biological function of X is a beneficial effect it has on some larger system that explains why X is there (as with the heart, which contributes to the fitness of whole organisms). But there is no need to look for too close a match with older teleological concepts. Colin Pittendrigh (1958) coined the term "teleonomic" for the relatives or descendants of teleological concepts that have a place within an evolutionary worldview; there is no need for a complete alignment of the teleonomic with the teleological. Another aspect of this issue concerns the degree of definiteness, sometimes called *determinancy*, of teleological facts. In some discussions of biological functions in philosophy the aim has been to make very fine discriminations—the function of this part of the frog's brain is to *detect flies*, not to *detect food* or to *detect small moving dark things*. But if descriptions of functions, in this sense, are summaries of evolutionary histories, there is no reason to expect this sort of sharpness.

Another issue in this area concerns the relation between quasi-teleological or teleonomic concepts and the normative or evaluative side of teleological thinking. In a traditional sense, the function of something is what it is *supposed* to do, in a sense such that things have gone *wrong* if something else happens. Within a minimal concept of function as causal role, or contribution to the activities of a system, this evaluative way of thinking about functions is clearly out of place, except as a kind of pretence. If something does not play its usual role in explaining activities of a more complex system, that need not be a bad or improper thing; it might be a step in the right direction for all concerned. In the case of the etiological concept of function, something's function is the effect it has that explains why it is there. This, again, brings with it no implication of goodness or propriety. The function has a role in explanations of a certain kind, and that is all. Descriptions using the word "function" can sometimes seem richer, stronger; there can seem to be a warm glow of purpose about a thing when it fulfils its function. But this is a holdover from earlier views. It does not correspond to anything in our present understanding of living systems.

FURTHER READING

On adaptation, Brandon (1990), Orzack and Sober (2001); on complexity and organization, McShea (1991); on construction, Levins and Lewontin (1985); on functions and teleology, Millikan (1984), Buller (1999), Huneman (2012).

Individuals

IN THE 1961 science fiction novel *Solaris*, by Stanislaw Lem, astronauts explore a planet where life exists, but does not seem to be divided up into discrete individuals. Or perhaps the oceanic planet is one big living individual. On earth, in contrast, living things seem to be conspicuously bounded, marked off from one another, and very numerous. In fact, if we think back to how the world would have looked in prehistoric times, living organisms would have been some of the *most* clearly bounded and easily counted objects, especially before people began making artifacts.

The obviousness, distinctive behaviors, and practical importance of organisms gave rise to what anthropologists call "folk biology," a set of habits of thinking about living things that all human cultures seem to share (Medin and Atran 1999). The obviousness of organisms also shows up in theories of all sorts. In the metaphysics of Aristotle, his main examples of "primary substances," the most basic things that exist, were individual horses and individual men.

In this informal, folk-biological sense, an organism seems to be something that does two things. An organism maintains itself—keeps itself alive—and reproduces, makes more things of the same kind. This is a useful way of thinking about life in many contexts, but as biology developed there was increasing recognition of puzzle cases—cases where there is certainly *life* present, but the living *thing* is less clear. The result was an ongoing discussion of "individuals" in biology, a discussion in which biological and philosophical questions are tightly entangled.

5.1. THE PROBLEM OF INDIVIDUALITY

The unearthing of problem cases began in the late 18th and early 19th centuries, especially in botany. Even familiar plants, such as an oak tree, raise problems. As small parts of a plant can often regenerate a whole, these parts seem to have a kind of autonomy. Perhaps the shoot or the bud is the true "vegetable individual," and a tree is a population of them. Further puzzles were posed by marine organisms such as corals and salps (T. H. Huxley 1852). Darwin, in the *Voyage of the Beagle*, puzzled over "compound" sea animals, where "the individuality of each is not completed" (1839, p. 128). Evolutionary theory soon transformed the discussion. Julian Huxley (grandson of T. H.) treated individuality as an evolutionary product, and saw the history of life as heading toward "the Perfect Individual" (1912, p. 3).

These fundamentals connect to more practical matters. As evolutionary biology developed it became more and more a *counting* science. How many offspring did this individual have? How big is this population? Counting is affected by assumptions about individuality—assumptions about when you have a new thing as opposed to more of the same. When the quantitative side of evolutionary theory was being worked out, people mostly thought about organisms where counting is easy, such as humans and fruit flies, but other cases are much less clear.

One recurring problem is the relation between *growth* and *reproduction*. Many plants make what *look*, at least, like new plants by growing them directly from the old. In "quaking aspen" (*Populus tremuloides*), what appear to be hundreds or thousands of trees scattered across many acres will be connected by a common root system from which they have grown (Mitton and Grant 1996). In the terminology used by John Harper (1977), there we have many *ramets*, but a single *genet*, or genetic individual. Similar phenomena are seen in violets and strawberries, which produce aboveground "runners" that give rise to new plants. In these cases the root systems are produced separately by each ramet, and it is easy for a runner to be broken, resulting in complete physiological separateness. Is this the growth of one continuing individual or reproduction by a single parent? Can we say whichever we like?

Maybe we should say different things in different contexts. Monozygotic human twins deserve two votes in elections, but perhaps they form a single unit in another sense.

A further set of problems is raised by "collective" entities—groups of living things that are in some ways like organisms or individuals in their own right. Important cases here include ant and bee colonies, and lichens. Each lichen is a close association between a fungus and many algae. Sometimes collective entities can clearly be living organisms in their own right; humans are collections of living cells. In other cases it seems that the collective should be treated as no more than an aggregation of lower-level individuals—consider a school of fish. Between the extremes there are intermediates. Some sea anemones form mat-like colonies, where there is some division of labor into reproductive forms and "warriors" that battle with other colonies, but where individuals interact only locally and the integration of the colony is very partial (Ayre and Grosberg 2005). A great many animals live in symbiotic partnerships with bacteria found on and within them, and these bacteria are often necessary for normal life in their larger partners.

One response to all this is to take a relaxed attitude. Perhaps a biological "individual" is just anything that some part of biology recognizes as worth describing. That is a reasonable view in many ways. But something is lost if we are *too* low-key about the issue. On earth, the distinctness of living organisms is a fact worth investigating. Biological objects recur, and persist as matter passes in and out of them. Evolution also from time to time creates *new* kinds of individuals—the eukaryotic cell, the multicellular organism, the ant colony. It's reasonable to look for a theory of how this works—an evolutionary theory of individuality.

5.2. DARWINIAN INDIVIDUALS

On an intuitive conception, living things are objects that maintain their organization, develop, and reproduce. I will start out with one of these, *reproduction*, picking up again the problem of distinguishing reproduction from growth. Some biologists, motivated especially by this problem in plants, have argued that what is called "asexual reproduction" in plants and other organisms is really growth, continuation of the same individual, because what is produced has the same genes as what was around before. An organism's unique genetic properties determine where it begins and ends.

In an elegant article called "What Are Dandelions and Aphids?" (1977), Daniel Janzen argued for a view of this kind. Both dandelions and aphids alternate between sexual and asexual "reproduction," where the asexual stage involves making an egg that is a genetic clone of the mother. Janzen argued that from an evolutionary point of view, a dandelion is a scattered object with many small parts that have each grown from these asexually produced eggs. An individual dandelion may be as big as an oak, though it has a very different shape; a dandelion is "a very large tree with no investment in trunk, major branches, or perennial roots" (p. 587; see also Cook 1980).

Whether or not it helps impose order on the unruly plants, this view cannot be applied in a general way. It has the consequence that bacteria do not reproduce when they divide (unless there is significant mutation in the process). Two strains of bacteria in a dish, one increasing in numbers because it can deal with a toxin that the other cannot, would not count as undergoing natural selection. A second problem with this view is the inevitability of *mosaicism* in multicellular organisms. Mosaicism is the presence of different genetic material, due to mutation and other forms of divergence, within a single organism. People often say the cells within a human are "genetically identical," but this is not literally true. We start our lives from one cell, but mutations accumulate with every cell division. Talk of genetic *identity* across a person's cells is an idealization; their cells are just very genetically *similar*.[1]

[1] This point is made dramatically by Austin Burt and Robert Trivers (2006): as there are about 10^{13} cells in a human body, 10^{12} cell divisions per day, and a mutation rate per cell division per nucleotide of about 10^{-9}, "this means every possible single nucleotide mutation occurs in our genome hundreds of times per day, and within our lifetime the whole range of Mendelian genetic diseases probably arises at one time or another, in one cell or another" (p. 421). To the extent that an organism is large and long-lived, it will be a genetic mosaic.

Let's start afresh. Reproduction is a product of evolution, as well as part of the evolutionary process. Reproduction takes different forms in different kinds of organisms—it is a different connection in different parts of the tree of life. Some forms of reproduction shade off into growth, and others shade into other things. Expressed simply, reproduction is the making of a *new* individual by one or more *parent* individuals, where the new individual is of the same kind, in a broad sense, as the parents. Complications arise with all parts of this formula—with the causal idea of "making," with the idea of "same kind," and, as we saw, with the boundaries between new and old individuals.

The varieties of reproduction can be divided into three different basic forms, and different problems arise around each. First, some things reproduce in a way that is entirely dependent on external machinery of some kind. Examples are viruses and genes. A virus can reproduce, but only by entering a cell and inducing the cell to copy its genetic material and make protein coats for new virus particles. A gene, similarly, cannot reproduce "under its own steam" in the way a cell can, but DNA molecules are copied by cells in a way that generates parent-offspring lineages of DNA molecules. Things like genes and viruses can be called *scaffolded* reproducers; they reproduce with the aid of much external machinery.[2] Cells, in contrast, do rely on external conditions, but the machinery of reproduction is internal to them. Things like cells can be called *simple* reproducers. Third, there are *collective* reproducers. These are reproducing objects that are made up of simple reproducers (or made up of smaller collectives, which in turn are made up of simple reproducers). There are no sharp boundaries between these categories. A eukaryotic cell, for example, has some features of a simple reproducer and some features of a collective reproducer, because the mitochondria it contains have remnants of a capacity to reproduce independently.

The three kinds of reproduction raise different problems of analysis. Here I will discuss just collective reproducers. These are cases where the question of distinguishing reproduction from

[2] This use of the idea of "scaffolding" is derived from a concept used by Sterelny (2003).

growth arises, and where interesting issues concerning colonies and societies are also seen.

One way to distinguish reproduction from growth is to look for a "bottleneck," a stage in the life cycle that reduces down to a single cell (Bonner 1974, Dawkins 1982). A bottleneck marks a new turn of the life cycle; the things on each side of it are different individuals whether they have different genotypes or not. This fits, in an intuitive way, the idea of reproduction as a "fresh start," and it is also important from an evolutionary perspective. Because a bottleneck forces the process of growth and development to begin anew, a small mutation in the initial stage can have a multitude of downstream effects. In Janzen's dandelions and aphids, the new objects produced do go through a one-celled bottleneck, so these are cases of reproduction. This is not an all-or-nothing matter, however. There can be partial narrowings in a life cycle, as well as narrowings to a one-celled stage. This is seen in aspens and strawberries making ramets through roots or runners. The bottleneck is not one cell wide, but it is narrower than what is to come. There is a partial fresh start.

Narrowings of this kind are also seen in cases of *metamorphosis*, which in many cases include the death of a majority of cells in the organism's body. Biologists have wrestled with the distinction between reproduction and metamorphosis (Bishop et al. 2006). Metamorphosis has an extra feature that distinguishes it from reproduction of an evolutionarily important kind, however, and that is the fact that in metamorphosis a "parent" can have only one "offspring"; there is no possibility of *multiplication* as opposed to mere replacement. When there is no multiplication in a population, the only way for there to be fitness differences is for the population to continually get smaller.

A second important feature of collective reproduction is the presence of a *germ line*, or some other form of reproductive specialization. In mammals like us, for example, only a small proportion of cells can give rise to a new whole organism; germ line cells are "sequestered" for the production of eggs and sperm. Our other "somatic" cells can reproduce as cells, but they cannot (by natural processes) give rise to a new human. In honey bee colonies the queen reproduces (along with the male drones), and the female workers do not. In many other insects, including other bees, there is no reproductive division of labor. This distinction helps mark a divide between cases where there is a group of insects (or cells, in our case) who happen to live and interact together, and cases where the colony (or organism, in our case) is a reproductive unit in its own right.

A third feature might be added. When we look at a bee colony and compare it to (say) a school of fish or a buffalo herd, another obvious difference is the overall level of integration and division of labor. Often the presence of a general division of labor is associated with a reproductive division of labor, but the two are not completely correlated, and perhaps both are important in their own right.

So three marks of genuine reproduction in collectives are the presence of a bottleneck, a germ/soma divide, and overall integration of the systems that reproduce. I see these as features that can be present in degrees. As a result, they can be mapped in a space, as in Figure 5.1. Here some different cases of collective reproduction are represented with respect to whether they have high, intermediate, or low "scores" (0, 1/2, or 1) on the three features. On the upper right are animals like us, where reproduction goes through a one-celled bottleneck, with a germ/soma distinction, and the reproducing unit is highly integrated. Oak trees differ from us in having much less germ/soma specialization. An aspen forming ramets is distinguished from the oak in not reproducing through such a narrow bottleneck. *Volvox carteri* is a green algae that forms colonies, where some cells function in swimming and others are specialized for reproduction (Kirk 1998). Each colony starts from a single cell and there is some overall integration of the system, but less than in an organism like us. In slime molds, in contrast, colonies form by the aggregation of many cells that forgo independent living in the soil to form a reproducing unit, but there is some germ/soma specialization. At the bottom left is a collective with low scores on all three features.[3] As I said, the

---

[3] This is a simplified version of a figure in Godfrey-Smith (2009), a work that contains more detail about the three parameters and the mapping of cases. The original figure was prepared by Eliza Jewett-Hall.

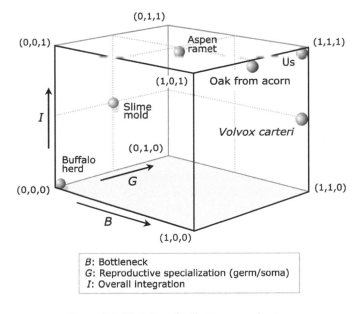

Figure 5.1. Varieties of collective reproduction.

three kinds of reproduction raise different problems. Bottlenecks, germ/soma distinctions, and overall integration are useful in dealing with collective reproduction, but don't seem to help with the other categories.

The previous three pages outlined my own framework for thinking about reproduction. Perhaps there is a better one. The more fundamental ideas here are that the biological world contains many modes of reproduction, and reproduction shades off into various other phenomena. There is a part-whole hierarchy of reproducing entities, and some activities of reproduction include reproduction by their parts. Looking again at our own case: a human cell reproduces by dividing, cell division includes the reproduction of a cell's genetic material, and the organized reproduction of many human cells is the reproduction of whole human beings. In any objects that reproduce, evolution can take place.

73

So cells, genes, organisms, and various other things are *Darwinian individuals*—things that take part in processes of evolution by natural selection.

These ideas further clarify the "units of selection" debates discussed in section 3.3. To ask whether something is a unit of selection—either in general or in a particular case—we should ask whether those entities vary, pass on traits in reproduction, and differ in reproductive success. The same test is applied to all cases, including genes, organisms, groups, species, artifacts, and ideas. For some of these objects it is hard to work out what reproduction involves, but that is what to look for. Once the situation has been clarified in this way, it is an empirical question which objects pass the test, and also which are units in significant evolutionary processes as opposed to minor or trivial ones. These questions arise especially for collectives, where there can be evolutionary processes at many levels at once. Consider a situation where organisms or cells are collected into distinct groups and reproduction occurs at two levels. The situation might be one where evolution within groups is very vigorous and leads to all sorts of new traits, accompanied by an occasional, less important process in which whole groups die out or split into two. It might be the opposite; it might be that the groups are all quite internally homogeneous, so there is little evolution within them, while a great deal of evolution goes on in the population *of* groups, with some groups reproducing more than others and passing traits to their offspring groups.

In 1995 John Maynard Smith and Eörs Szathmáry published *The Major Transitions in Evolution*, a book that tried to explain a small number of landmark events in the history of life, including the origin of life itself, the evolution of the cell, the evolution of sex, the evolution of multicellular organisms, and the evolution of language. Maynard Smith and Szathmáry saw many of these "transitions" as inventions of new ways of passing information across generations, an idea I will look at in the last chapter. Whether information is central here or not, many of these transitions are events in which new kinds of Darwinian individuals arise from old. New kinds of objects become able to reproduce, form parent-offspring lineages, and undergo evolution in their own right. They are "transitions in individuality" (Michod 1999). One example is the evolution of the eukaryotic cell, a process initiated by the swallowing of one cell by another, perhaps 1.5 billion years ago. The descendants of the swallowed cell (or cells) include our mitochondria. Another example is the evolution of the multicellular organism. This happened several times, probably in each case by a cell dividing in such a way that its "daughter" cells did not separate but stayed attached. In many cases the resulting unit never evolved much complexity, and its descendants live on, if at all, as thin weedy filaments in the sea. But in other cases the results led to the evolution of animals like us.

These are all events in which new Darwinian individuals arise, and also events in which the evolution of new Darwinian individuals leads to the partial suppression of old ones. In the evolution of the eukaryotic cell and the multicellular organism, lower-level entities become partly *de-Darwinized* by the evolution of the new unit. What I mean is that they lose—in part—the features that give rise to a significant Darwinian process. The cells in our bodies are an example. These cells vary, reproduce, and inherit traits from their parent cells. They are still Darwinian individuals, but their evolutionary activities have been largely curtailed, and this happens as a result of the evolution of features that I earlier said were marks of genuine reproduction at the collective level, the level of multicellular organisms. Multicellular collectives like us have, in effect, moved through the space seen in Figure 5.1, becoming clearer cases of reproducing entities in their own right, and their movement has consequences for the evolutionary capacities of their parts. The cells within a single human body are genetically very similar to each other, as they are all derived from a one-celled zygote (the bottleneck). And whatever advantage one cell might gain over another within an organism has little long-term effect unless these cells are in the germ line. What matters instead is the survival and reproduction of large colonies of these cells, also known as human beings. The evolution of multicellular organisms has partly de-Darwinized the cells that gave rise to them.

5.3. LIVING THINGS

The previous section was about reproduction, especially its role in evolution. The chapter, though, began with *organisms*. How do they fit in to the story?

Reproduction is part of the intuitive or folk-biological view of organisms, but being a Darwinian individual is not the same as being an organism. Some Darwinian individuals are not organisms; examples include genes and chromosomes. These objects are reproduced in a way that is evolutionarily important, but they do not reproduce with their own machinery (they are scaffolded reproducers). Viruses are a more controversial example. They have more independence than genes or chromosomes, but can do nothing without the metabolic capacities of cells.

A picture emerges: once there are organisms, which control energy and the machinery of reproduction, other things can be reproduced *by* organisms. So the Darwinian individual category is wider than the organism category. Furthermore, collections of organisms sometimes come to work so closely together that they can reproduce *as* groups or colonies. Eusocial insects are examples. Some people see these as organisms in their own right (Hölldobler and Wilson 2008), but even if they are not organisms, they can still be Darwinian individuals.

How about the other possibility: are there organisms that are not Darwinian individuals? Initially it seems that this won't happen: evolution is how organisms come to be (unless there is a divine creator), and as all organisms will be part of an evolutionary process, they will be able to reproduce. But that argument goes too quickly—there are other ways things can fit together.

At this point we need to take a closer look at what is meant by "organism." In some interpretations, being an organism necessarily requires being able to reproduce, or perhaps being the sort of thing that can reproduce. But even if this is one sense of the term, there is room for other views, and for a category that does not tie being an organism so closely to reproduction. This is a *metabolic* view of organisms: organisms are systems comprising diverse parts that work together to maintain the system's structure, despite turnover of material, by making use of sources of energy and other resources from their environment.

This view can be challenged in several ways. Many formulations are too vague to deal with hard cases, and they need at least to be sharpened up (Pradeu 2010). But I will work within this approach without settling all the details. On this conception, an organism can have any history, in principle, and reproduction is optional. An organism might persist indefinitely without making more organisms. Organisms are essentially things that persist, using energy to resist forces of decay and maintain their distinctness from their surroundings, and only contingently things that reproduce.

Within this framework, it's then possible to argue that there are organisms, perhaps many of them, that are not Darwinian individuals. This is because of an argument about *symbiosis* (Dupré and O'Malley 2009, Pradeu 2012).

Most or all plants and animals live in close association with symbionts, especially bacteria present within and on them. There are more bacterial cells in your gut, for example, than there are animal cells in your entire body. These bacteria have an important role in metabolism and development. Sometimes microbial partners of this kind are transmitted "vertically," between host parent and host offspring, as part of reproduction. An example is the bacteria that aphids (making a second appearance in this chapter) have inside them that make it possible for an aphid to live off plant sap. But other symbionts are acquired "horizontally," from various sources in the environment. In some of these cases, it is possible to make the following argument: the organism—the metabolic unit—is a system comprising a familiar animal (e.g., a human) *plus* its microbial symbionts. This argument can be made by noting the metabolic integration of the partners, how they help each other stay alive, and it can also be made, at least in some cases, by noting that one or both partners *tolerates* the other with respect to its immune responses. Pradeu (2012) argues that immune responses can be used quite generally to mark out where organisms begin and end.

So an organism, perhaps, can comprise a collection of animal cells plus a collection of microbes acquired from its environment.

In cases where the microbes within an animal are acquired from the wider environment rather than from the animal's mother, the resulting "consortium" does not reproduce as a unit. The host animal and the microbes are each part of their own parent-offspring lineages, but the combinations are not. Then these "consortia" are organisms but not Darwinian individuals; they are products of the joint action and joint evolution of two or more kinds of Darwinian individual, which come together afresh in each generation. This argument can be applied to a great many animals, including ourselves. (Some of our symbiotic microbes come from our parents but many do not).[4] The surprising idea here is that it is *not* true that a typical organism is a metabolic whole that also reproduces as a unit.

Let's consider more closely the idea that metabolism, the use of energy to maintain organization, is central here and some tightly bound symbiotic combinations are organisms because normal metabolism requires both partners. This line of reasoning can lead to some strange places. It looks OK when one partner lives inside the other, but what if two metabolically integrated partners live at some distance, each making use of the products of the other? Is *that* "consortium" an organism? If so, what about ourselves and all the photosynthetic organisms making the oxygen we need to stay alive? Where does this stop?

Maybe it "stops" nowhere, and we have made a mistake to think of life as a feature of living *things*, definite objects separated one from another in space. Rather, living activity is a more spread-out affair, one in which a range of physical parts interact to maintain metabolic patterns. The "Gaia hypothesis," the idea that the whole earth is a living organism (Lovelock 2000), is an extreme version of this idea, but it need not be defended in this extreme way. More and more distant factors become less and less metabolically important to any given biological object. A case of living activity might be almost entirely localized to a tiny film of water, even though metabolism within that system is the joint

---

[4] For a description of the diversity of human internal microbial communities and their origins, see Ursell et al. (2012). The issues in this section are discussed in more detail in Godfrey-Smith (2013).

product of many objects. In a collection of reproducing parts as large as the whole earth, there is no reason for the parts to cooperate, and there will be many opportunities for one part to exploit others. On a smaller scale, where the partners are more tightly associated, cooperation between very different partners can be viable, and can include tight metabolic connection.

An alternative way to approach these cases is to draw on the replicator/interactor framework discussed in section 3.3. This framework was developed as a general way of thinking about the objects that figure in evolution. Replicators are copied faithfully, and interactors are (usually) larger objects that are constructed by replicators and assist their replication. In chapter 3, I rejected this approach because it is a mistake to say that replication is necessary for evolution by natural selection. But the other part of the framework, the idea of an *interactor* as an evolved object, might be useful in dealing with symbioses and the like (Sterelny 2011). There are objects that recur in evolution without reproducing as units. Their parts reproduce, and the parts come together to make more of these recurring objects. Looser symbioses are easy to see in this way. For example, some shrimp and small fish form associations and live, apparently harmoniously, in the same den. (Often a pair can be seen poking their heads out of a hole together.) Some acacia trees build hollow structures that house ants that guard the tree, and in some cases the trees also feed the ants. A tree-plus-ant colony does not reproduce as a unit; these combinations arise when new ants and new trees come together. Perhaps human beings are interactors in the same sense.

I will look at one more topic to finish this chapter. Something you might have expected at the beginning of a philosophy of biology book is a section called "What Is Life?" But the topic belongs here, now that some ideas have been laid out.

Modern biology has *partly answered* and *partly deflated* the question of the nature of life. Saying this does not depend on the more speculative ideas in this section; the point is general. The "deflation" of this issue is evident especially in contrast to how things looked in the 19th century. During that time the mechanistic project in biology developed. As it matured, the obstacles it faced became clearer. Life appeared to be very distinctive, possibly an addition to the physical-mechanical universe. How do things look now? We have a fairly good understanding of all the activities that go on in a living organism (except for experience and consciousness). We know how metabolism works, how organisms use matter and energy to maintain their organization. We know how reproduction and development work, and how organisms evolve. Once those topics have all been tackled, the appearance of a single special property—*life*—fades away. Our theories explain why metabolism, development, and reproduction are mostly present in the same objects: metabolism arises through evolution, reproduction mostly requires the metabolic control of energy, and a living thing usually has to develop before it can reproduce. But we can also see why some of these features can be present without the others. It makes sense that viruses exist, for example, entities that reproduce despite not having a metabolism. (If viruses had not been discovered by now, it would make sense to predict them.) Theories of evolution, development, reproduction, and metabolism cover everything you might want in a theory of life, but life itself partly recedes from the scene.

FURTHER READING

For reproduction, Griesemer (2000, 2005), Blute (2007); for evolutionary transitions, Buss (1987), Calcott and Sterelny (2011); for organisms and individuals, Santelices (1999), Pepper and Herron (2008), Queller and Strassman (2009), Folse and Roughgarden (2010), Bouchard and Huneman (2013); for puzzle cases, especially plants, Bouchard (2008), Clark (2011); for life, Bedau (2007), Dupré (2012).

Genes

THE FIRST CHAPTER gave an overview of some of the history of biology, and this chapter begins with a closer look at one area, genetics. The history feeds directly into a central issue, the question of what genes *are*. The following section looks at what genes *do*, and the last looks at their role in evolutionary processes.

6.1. THE DEVELOPMENT OF GENETICS

In the 1850s and 1860s the monk Gregor Mendel did a series of experiments in plant breeding that led him to the postulation of inherited "factors" affecting traits of organisms. One experiment began with two lines of pea plants, a line that produced only purple flowers and another that produced only white flowers. Mendel crossed them, in sexual reproduction, and found that all the offspring had purple flowers. Those were allowed to self-fertilize, and in the next generation he found a three to one ratio of plants with purple flowers to plants with white. So white disappeared but then reappeared. This suggests that white flowers are due to an inherited factor that can be masked and revealed—something passed intact through the apparent loss.

Results like this suggest a hypothesis: each organism contains a pair of *factors* affecting each inherited trait, receiving one factor from each parent. When sex cells are formed in the new generation, there is an equal chance of the sex cell containing the factor from one parent or the other. (See the discussion of "Mendel's First Law" in §2.1.) An organism's observable properties arise from their combination of inherited factors.

Mendel's work was largely ignored till about 1900, when these phenomena were reproduced and the ideas rapidly advanced, especially at the hands of William Bateson (1900). Mendel's factors became known as "genes," a term due to Wilhelm Johannsen (1909). Johannsen also introduced the distinction between *genotype* and *phenotype*. An organism's phenotype is all its observable characteristics, due to both genes and other causes; the genotype is the organism's underlying genetic nature. At this stage the only access biologists had to genes was through breeding experiments; genes were hypothetical entities used in explanations of certain differences between organisms. Where there are two versions of a characteristic—purple or white, tall or short—that are inherited in a Mendelian pattern, this shows there are two *alleles* (alternative forms of a gene) affecting that character.

Thomas Morgan and his group, working especially in the 1910s and 1920s and introducing the fruit fly as a model organism, achieved results that appeared to give genes more materiality. Genes lie on chromosomes, with some close to each other on the same chromosome, some distant, and some on different chromosomes. This explains why some genes at different *loci* tend to be passed on together, while others are passed on independently. In his 1933 Nobel Prize lecture, however, Morgan insisted that genes could still be fictions, not material things; it did not matter. How could a gene be "referred to a definite location in a chromosome" but still a mere fiction? This early step illustrates the odd road that genes have taken as objects.

H. J. Muller's work, which included discovering the role of X rays in causing mutations, shifted attention more squarely to the chemical nature of the gene. Beadle and Tatum (1941) developed the "one gene-one enzyme" hypothesis: what a single gene does is make an *enzyme*, a protein that controls a reaction in the cell. Traits of organisms can be affected by many genes, but that is because they are affected by many enzymes. At the end of this period of "classical" genetics, genes were seen as unknown physical things that are linearly arranged on chromosomes, passed on in the ways described by Mendel, and each responsible for making one enzyme.

The 1950s see a transition to "molecular" genetics. A series of discoveries about the chemistry of cells culminated in the 1953 Watson and Crick model of DNA. Through the 1960s the "genetic code"—the mapping between a DNA sequence and the structure of a protein—was uncovered. The resulting picture had it that DNA acts as a template, through an RNA intermediate, in specifying the linear order of amino acids within protein molecules. The first stage is *transcription* of DNA to mRNA, followed by *translation* of RNA into protein. Proteins then form complex three-dimensional shapes (more or less spontaneously, though in a way affected by molecules around them), and this produces the molecules that do most of the work in cells. The 1960s also saw the first work on the regulation of gene action. The expression of genes is controlled by molecules binding to the chromosome to promote or inhibit transcription of the DNA (Jacob and Monod 1961).

If that is what DNA does, what is a gene? Here ideas were carried over from earlier work, especially the "one gene–one enzyme" view. Seymour Benzer (1957) introduced the term "cistron" for the unit of function in DNA, and the cistron came to be identified with a stretch of DNA that specifies the structure of one protein molecule. A cistron is accompanied by regulatory regions—stretches of DNA near the cistron that affect the cistron's transcription into mRNA.

This is *roughly* a vindication of classical genetics: genes were introduced as objects that play a causal role, and later work uncovered what actually plays that role. Genes, it could be said, turned out to be stretches of DNA that code for proteins. This is sometimes called the "neo-classical gene" (Portin 2002). There were initial mismatches, though, between the classical and molecular pictures, and a gradual accumulation of these followed—a continuing series of discoveries that are partly at odds with the classical picture.

Why was it necessary to introduce a term like "cistron"? The reason was to overcome ambiguities, especially due to the *recombination* of genes. Recombination occurs by the shuffling of whole chromosomes into new combinations through sex, and by *crossing over*—the mixing of material from two chromosomes during meiosis, the form of cell division that makes sex cells. In crossing over, two chromosomes break and exchange material. This process does not pay attention to the boundaries between cistrons. The only unit that cannot be split is the single nucleotide (with its four forms C, A, T, and G). The unit that is *passed on intact* is not the same as the *unit that makes a protein*. Later work discovered that the parts of the genome that make a single protein are often not localized, contiguous pieces. In eukaryotes like ourselves, coding stretches of DNA are broken up by noncoding stretches, *introns*. These are transcribed into RNA but edited out before proteins are made. Still further work found that "raw" RNA transcripts are often processed into many different finished transcripts for protein synthesis. Sometimes two RNA transcripts derived from different chromosomes are tied together to yield a finished mRNA (RNA *trans*-splicing). The DNA coding for one protein can also overlap with, or be embedded in, the DNA coding for another.

The importance of gene regulation in explanations of genetic phenomena also grew. A typical gene has a *promoter* near it on the chromosome, to which molecules bind which facilitate or inhibit transcription, and the gene may be affected by other regulatory regions further away. Some definitions of "gene" include regulatory regions as genes in themselves, some include them as part of the genes they affect, and some do not treat them as genes at all (Waters 1994). Regardless of how questions of definition are handled, a lot of noticeable differences between organisms, including famous cases in classical genetics, have turned out to involve not a difference between the proteins being coded for— different sequences in a cistron—but differences in regulatory regions. In what I think is something of an irony, the "white-eye" mutation in fruit flies that got T. H. Morgan's research program moving, an intensely studied classical gene, turned out to be a mutation in a promoter. A retrotransposon (§6.3) inserted into the promoter and inactivated it, so the cistron was not transcribed.

How much of a disruption is this of earlier views? I spoke above of a series of discoveries that are *partly* at odds with the classical picture. You can emphasize the match between the two, or the mismatch, and one or the other might seem more important depending on what is being discussed. One approach is to distinguish several senses of the term "gene," useful in different contexts. A long-standing duality here, which we saw above in Morgan himself, is between using the word "gene" merely to organize talk about observable differences between organisms that show up in breeding experiments in certain ways, and as an attempt to refer to a real hidden object of some kind. Along related lines, Lenny Moss (2002) distinguishes what he calls the *gene-P* and *gene-D*. A gene-P (P for predictive, and for preformation) is anything in the genome that has a predictive role in relation to a certain phenotype. So a *gene for breast cancer* is anything in the genome that tends to predict breast cancer in otherwise normal circumstances. A gene-D (D for development) is a region of DNA that acts as a template in the synthesis of a gene-product (protein or RNA). Others recognize three or four distinct concepts (Griffiths and Stotz 2013). In an earlier discussion, Sterelny and Griffiths (1998) offered a more minimal view— almost a non-analysis—saying that the term "gene" has become a "floating label" for *any* bit of DNA that is of interest. Perhaps *any* is too strong, but there is an element of truth here (and in later parts of this book I use the term "gene" with a fair bit of this buoyancy).

Setting aside questions about terms, what matters here is that it seemed for a time that a single, definite kind of thing played a certain role, and that unity partly dissolved. To some extent the classical gene has had its nature explained, to some extent the concept has been augmented, and to some extent it has been replaced. Suppose we were starting from scratch, with the finer-grained information we have now, and had not gone via Mendel and Morgan. Would we talk about genes, as units, at all? What cells contain is *genetic material*, and different stretches and chunks of it play different roles. Depending on the inheritance system, different pieces of genetic material can be reliably inherited, and depending on the context, different pieces of it can play causal roles in cells and in organisms. The postulation of Mendelian "factors" and then particle-like genes was enormously productive in the early 20th century. Seeing the gene as an atom of inheritance led to decisive progress. Progress 100 years later is coming from seeing it differently.

6.2. GENE ACTION

This section is about what genes *do*—about their causal role. This area of debate is especially intense because of the specter of "genetic determinism" about human characteristics, and other versions of the idea that genes are the overriding causes of what any organism is like. On the other extreme are taken to be *tabula rasa* (blank slate) views, which give the environment a similar primacy and downplay genetics. In recent discussion the problem has often been to avoid these extremes while asserting more than a bland interactionism—a view saying merely that every feature of every organism is due both to its genes and environment and there is no way to distinguish their importance.

At this point it is time to take a look at the idea of causation. The concept of cause is a bit of a mess, pulled about in different directions.[1] But some headway can be made by noting that there seems to be a duality in causal thinking; there are two sets of criteria that guide causal claims, and two relevant relations that one thing can have to another. First, a cause can usually be thought of as a *difference-maker*. This is seen in reasoning of this form: C caused E because if not for C, E would not have happened. Second, there is a way of thinking about causation in which a cause is something that *produces* its effect, by some local connection between the two. In pulling these criteria apart, philosophers make use of special cases such as *redundant* causation and causation by *omission*. Redundant causation: *C1* produced E, but *C2* was ready as a backup if *C1* had failed. (Think about a team of assassins.) Then although *C1* produced the effect, it was not a difference-maker. Causation by omission: something can affect an outcome by *not* interacting physically with the chain of events leading to it. You can often be a difference-maker (for example, in a meeting) by keeping quiet and doing nothing.

Recent work on causation has focused especially on difference-making, and has developed a sophisticated framework based on the idea of *intervention*. If the state of X is a cause of the state of

[1] In Godfrey-Smith (2010) I discuss these issues about causation in more detail.

Y, then you can change the state of Y by manipulating X, where this manipulation holds fixed other factors "upstream" of Y, except for factors that lie on a path between X and Y (Pearl 2000, Woodward 2003). This is not a ground-up analysis of causal facts, because it takes for granted the idea of manipulation and the idea of a "path" I appealed to above. Also, these manipulations are often not possible in practice, and are merely hypothetical. So the whole framework has contentious elements. Its value lies in the way it can clarify causal description in complex systems and elucidate how causal facts are inferred from empirical data.

This framework links to some older ideas within biology, and the two together provide a way of thinking about the relation between genes and phenotypes. The *norm of reaction* for a genotype is a function (in the mathematical sense) from environments to phenotype that is characteristic of that genotype (Schmalhausen 1949, Schlichting and Pigliucci 1998). (It is really not "an environment" but a sequence of developmental environments that usually matters, but I will ignore that here.) A genotype shows *phenotypic plasticity* when the resulting phenotype is very sensitive to the environment.

There will also be a function for a given environment, *from* different genotypes *to* phenotyes. For any phenotypic trait, we can then ask what sort of variation it is sensitive to and what it is insensitive to. Conrad Waddington (1942) coined the term "canalized" for a phenotype that appears reliably in individuals of a given species across significant variation in environment and in genotype. The term is often used now to refer just to stability in the face of environmental variation, but the same questions about sensitivity of outcome can be asked about both genes and environment.

This framework can be connected to the interventionist approach to causation. Just about all traits will be products of both environmental and genetic causes. But the *kind* of influence will vary. Some traits "do not notice" a good deal of variation in the environment; manipulating the environment is not a very effective way of altering them. Or the only variation in the trait that can be brought about by altering the environment would give you a dead organism, not a live organism with different features. Among the difference-makers affecting organisms are some with high *specificity* and others with less. *X* is a *specific difference-maker* for *Y* if variation in *X* over many different values leads to variation in *Y* over many different values.[2] Both the words on a newspaper page and the presence of oxygen in the air are difference-makers with respect to what you come to believe when you read that newspaper, but the words on the page are a more specific difference-maker than the presence of oxygen in the air. (Well, maybe this should be tested, but the principle is clear.)

With this framework we can state obvious facts and set up further questions clearly. Environment is more important than genetics for which particular language you speak, even though a lot of genes need to be in place for you to speak that language. Many debates about the causation of human characteristics like intelligence and personality are debates about which factors are specific difference-makers, and which are genuine causes without being specific difference-makers. The specificity of a difference-maker comes in many degrees and varieties. This framework avoids what I called "bland interactionism": genes and environment both affect every trait, but there are coherent ways to distinguish their roles based on difference-making and specificity.

When the aim is causal explanation at the level of whole-organism phenotypes, this focus on difference-making provides the best framework presently available. Let's now go down to the level of the cell.

DNA has several overlapping roles within a cell.[3] It is a template used in the manufacture of protein molecules, via RNA intermediates, and it is part of a control system that regulates which proteins are made in which cells at which times. It has this second role both by making gene products that act in control processes, and by containing sites at which molecules that affect the expression of the cell's DNA can bind. (DNA is also used as a template for RNA molecules that are not directly on the path to proteins,

[2] This framework is developed and applied to genetics in detail by Waters (2007) and Griffiths and Stotz (2013). Lewis (2000) developed related ideas about degrees of influence. See also Woodward (2006).

[3] Some of what I say about the role of gene action in cells here has exceptions in viruses, which I discuss occasionally, and also in many fungi, in which cellular organization is very partial.

such as rRNA, and it is replicated in the process of cell division.) In classical genetics a gene is a cause; it is at the beginning of a chain by which we explain something about an organism. Once genes are seen as embedded in causal networks, subject to control by signals and other factors, "gene action" becomes an effect as well as a cause.

Genes are difference-makers for many events that happen within a cell. At the cellular level, though, more can be said: genes are difference-makers *because* they are templates for other molecules and binding sites for regulatory proteins. Enzymes, in contrast, are difference-makers because they catalyze chemical reactions, while DNA does not do this. I said earlier that causal thinking is guided both by the idea of difference-making and also by the idea of the production of effects by causes. The points made just above seem to be leading toward a general picture: perhaps causal description of a system at a low level, with respect to parts and wholes, is about the production of one event by another, and once we zoom out to a level at which the mechanistic detail has disappeared, what we see is difference-making. That is not quite right, however, because within the mechanisms of the cell, described from very close-up, a lot of what happens is due to things *not* acting when they might act, to one thing preventing another, and so on. Those are the marks of difference-making as a distinctive kind of causal relation. So the situation is more like this: in an organized system of interacting factors like a cell, a multitude of facts about the local production of one event by another give rise to a further set of facts about difference-making in that system. The way this happens depends on the spatial arrangement of the system at the relevant time, as well as the activities of those parts. The arrangement of the parts enables some factors to be difference-makers by *not* acting, and also enables one event to produce another *without* being a difference-maker, because backup causes are available. Here, and more generally, production plus spatial organization is the basis for difference-making.

For many biologists, what has been said so far—about networks, difference-making, and so on—might be true enough, but it leaves out some important facts. First, while it's true that a gene can't be transcribed without polymerase enzymes, will have its RNA products edited by other enzymes, and so on, *those* molecules are gene products too; they must have been made earlier by the genome. And if the external environment has an influence, the receptors that mediate those influences are products of the genes. A single gene can't do much on its own, but—many will want to say—what we have been calling "other factors" are products of the genome as a whole, if we go one step further back.

Is it really true that the genome at time t_1 makes the factors that affect gene expression at time t_2? No, the whole cell at t_1 does this, including various molecules, cellular structures, and signals from outside. The cell and its environment at t_1 give rise to the cell at t_2, including gene action at t_2. The entire machinery of the cell at one time (in conjunction with external factors) gives rise to the entire machinery of the cell at a later time. Merely looking further back in time does not change things. But a related argument can then be raised: while cellular activities and the phenotype can be affected by many things, perhaps genes have a special role because they *program* the development of the organism, or *code for* their effects, while other factors do not.

This brings us into a jungle of questions about information, codes, and programming in biology. I will discuss these issues as they relate to genetics here, and return to them in the last chapter, where many parts of biology that encounter the concept of information are considered at once.

The original and narrow sense of the "genetic code" was a mapping from nucleotide triplets to amino acids, the mapping by which DNA specifies the structure of protein molecules (Crick 1958, 1970). This is indeed a code-like relationship, one that has features in common with language. A small alphabet of elements is used to form many combinations, and a kind of "reading" of the nucleic acid sequence, following a fixed rule, occurs at the ribosomes where mRNA is used to construct a protein.[4] The existence of this relationship does not support the idea that genes code for the phenotype of the whole organism—number of fingers, intel-

---

[4]For detailed arguments on this point see Godfrey-Smith (2000), Griffiths (2001).

ligence, being able to learn a language—or anything downstream of a protein molecule. It can be convenient to use "coding for" talk as a shorthand, as a variant on the "gene for" shorthand that is used to tie genes to phenotypic characters they are reliably associated with. This can be misleading though, as it can make it seem that there is a special kind of *instructing* of the eventual phenotype going on, one that reaches through the contingencies of the network connecting the two.

One significant difference between genetic and non-genetic causes is that a gene can have an effect on the phenotype that explains why the gene is there—why it has been selected for. Like any other part of an organism, a gene can have a biological function to bring something about (§4.3). Non-genetic causes of the phenotype do not usually have this feature, though they may in special cases.[5] If a gene has a function in this sense, that is a fact about the past, about evolutionary history; it does not enable the gene to push a bit harder than, or push in a different way from, other factors affecting how the organism turns out.

Another link between genetics and information technology involves the idea of a genetic *program*. The idea of a "program" in the genome has been used more loosely, in a less theoretically anchored way, than the idea of a genetic code. Some talk of a "program" is just a gesture toward the orderliness of biological development, and its being a product of evolutionary design. Then the idea of a program does not explain anything about *how* development works. It is not obvious how to make the idea of a genetic program or computation more substantial—so that it rules out some hypotheses as well as ruling some in. It is sometimes said that computation is a physical process that mediates between an input and an output in a way that mirrors logical or mathematical relationships. This view has the consequence that any physical process will count as computing *something*, a consequence that may or may not be a problem. There is a narrower category of physical processes, though, the ones that occur the way they do *because* they mirror logical or mathematical relationships; some process of selection or design has set them up that

[5]See Sterelny et al. (1996) and papers in Oyama et al. (2001).

way. The computers we build are in this category, and so are some parts of evolved genetic systems. Whereas the importance of the "code" idea was in the explanation of protein synthesis, the importance of the "program" idea lies in gene regulation. There are quite rich mappings between logical operations and the processes that govern gene regulation (Istrail et al. 2007). Gene regulation involves cascades of events in which one gene's products activate or inactivate transcription of another, which works with a third to regulate the behavior of a fourth, and so on. These cascades feature *and* and *or* and *not* operations; gene *A* might be activated if either *B* or *C* are active, or only if *B* but not *C* is active, and so on. Silencing a silencer of a gene is like double negation. Recent work in genetics has uncovered a degree of complexity in these switching operations that is far beyond what had been anticipated even a decade or so ago; much of what had been seen as "junk" DNA in human cells, for example, now appears to play a role of this kind (Djebali et al. 2012). To the extent that gene regulation involves a cascade of interactions that can be described in these terms, it is more than a vague metaphor to say that it is the execution of a program by the cell.

Earlier in this section I distinguished several roles DNA has within a cell. It is a template used in the manufacture of protein molecules, and it is part of a control system that regulates when and where these proteins are made. These roles can be linked to the two well-justified applications of information-related concepts in genetics isolated above: DNA codes for proteins, and computation-like processes within the cell control gene regulation.

This pair of roles was glimpsed quite early. The cell biologist David Nanney, in a farsighted quote from 1958, after Watson and Crick's model of DNA but before any of the genetic code had been worked out, said that two kinds of things are achieved in a cell:

> On the one hand, the maintenance of a "library of specificities," both expressed and unexpressed, is accomplished by a template replicating mechanism. On the other hand, auxiliary mechanisms with different principles of operation are involved in determining which specificities are to be expressed in any particular cell. . . . [These] will be referred

to as "genetic systems" and "epigenetic systems." (Nanney 1958, p. 712)

This breakdown of tasks can be expressed with the restricted use of the language of information technology defended above. DNA functions first as a cell-level memory, a "library of specificities," and its second role is as part of a control system, which often has a partly computational nature, that governs the use of this library. The control system itself is the whole cell, including both DNA and other structures that the cell continually rebuilds and some-times replicates. DNA is both a resource for the control system, its memory, and part of the controller itself. It is a library that has had other roles grow up upon it.

6.3. GENES AND EVOLUTION

Genes are treated as central in many theoretical representations of evolution, especially in formal models. Evolution has often been *defined* as "change in gene frequency." In chapter 3 I intro duced views that treat all evolution as competition between rival genes (Williams 1966, Dawkins 1976), and in chapter 5 I briefly treated genes as one kind of Darwinian individual. In this section I look at genes and evolution once again, drawing on the ideas outlined above in this chapter.

DNA is copied when cells divide, and this is reproduction in the sense relevant to an evolutionary process. As a population of organisms evolves, so too does their DNA. In the simplest cases, evolution in genomes and organisms is closely coupled together. Consider a population of cells reproducing asexually, by cell di-vision, with no exchange of genetic material. A mutation arises that gives one cell an advantage—the mutation modifies a pro-tein in a way that makes it able to detoxify a formerly destructive chemical. So the cell lives and reproduces when others die, and its descendants take over the population. Organisms with the muta-tion have higher fitness than their rivals, and so, in a way, does the mutated gene itself. Given the background in which it oper-ates, it does something that enables it to proliferate and spread.

93

What spreads is not just that bit of DNA, though, but the whole genome. The bit of DNA that has the good effects is not passed on as an evolutionary unit in its own right.

Now introduce the exchange of genetic material (through bacterial conjugation, eukaryotic sex, or something else). Once there is a mixing of genetic material, the evolutionary role of genes changes. A good new mutation can easily find itself in many different genetic backgrounds; it has an evolutionary path of its own.

People sometimes say that a gene is passed on as a discrete unit. Here is Richard Dawkins, in *River Out of Eden* (1995, p. 5): "Genes themselves do not blend. Only their effects do. The genes themselves have a flintlike integrity." This is not really what happens; it is an idealization. Suppose we are dealing with a sexual population like our own. As genetic material is copied in the production of sex cells, chromosomes are broken and parts of one are combined with parts of another. This process, as noted earlier, does not pay attention to the boundaries between cistrons, or any other units at a gene-like scale. The only unit that cannot be broken is the individual nucleotide. But small and medium-sized stretches of DNA can persist for long periods in the form of copies in this process, and a small stretch that does something useful for the organisms that house it may spread through the population, becoming more common than alternative sequences at that place in the genome. Thus an *allele* can *increase in frequency*. That is the kind of thing that standard models of evolution focus on, and describe in mathematical form. These models make it possible to see much of evolution as a competition between alleles for representation in the gene pool.

Suppose an explanation is given of the increase in frequency of some allele, some bit of DNA, in terms of the allele's own fitness. One argument that has been made holds that descriptions like this make for convenient accounting but cannot *explain* what happens, except in some special cases. Usually a gene does not have a consistent enough causal role for its dynamics in the population to be explained in terms of the gene's own fitness.[6] A gene

---

[6] For versions of this argument see Wimsatt (1980), Sober (1984), Lloyd (1988), Gould (2002).

has a role that can lead to evolutionary success or failure only in the context of other genes. But working within the particle-like model it is possible to assign fitnesses to genes in ways that accommodate these facts. Each allele may have a range of *context-sensitive fitnesses*. As John Maynard Smith (1987) argued, this way of thinking about context-sensitive fitnesses is no different from standard ways of thinking about the fitnesses of strategies in game-theoretic scenarios, such as the hawk/dove game (§2.3). The fitness of the hawk strategy is high in the context of a dove, low when the other partner is a hawk. If we know the frequencies of the contexts or backgrounds in which the strategies find themselves, we can work out their average fitnesses and hence which one will increase in the population. If explanations in terms of context-sensitive fitnesses are OK here, why not at the level of alleles?

BOX 6.1. A GENETIC MODEL OF EVOLUTION

Genetic models of evolution treat genes as particles that sex brings together into combinations. Suppose there are two alleles, A and a, found at a genetic locus and their frequencies are p and q respectively, where $p + q = 1$. Assume the population is diploid, mates randomly, and the generations do not overlap. Everyone reproduces at the same time and the same amount, if they live long enough to reach that stage, and they die after reproduction. Given random mating, the frequencies of the diploid genotypes AA, Aa, and aa at the beginning of a new generation will be p^2, $2pq$, and q^2, respectively. Suppose that genotypes differ in their ability to survive long enough to reproduce, however, and these fitness differences can be represented with fixed numbers W_{AA}, W_{Aa}, and W_{aa}. The frequency of the AA type after selection has occurred, at the time of mating, is $p^2 W_{AA} / \overline{W}$, that of the Aa type is $2pq W_{Aa} / \overline{W}$, and that of the aa type is $q^2 W_{aa} / \overline{W}$, where $\overline{W}$ is the mean fitness in the population, defined as $\overline{W} = p^2 W_{AA} + 2pq W_{Aa} + q^2 W_{aa}$. The frequency of the A allele in the next generation, p', will be $p(pW_{AA} + qW_{Aa}) / \overline{W}$ (compare Box 3.1).

Here alleles are the things being passed on, but diploid genotypes are the things with definite fitness values, as it is the whole organism that lives or dies. The average fitness of an allele can be worked out by taking into account the frequencies of the contexts the allele might find itself in: $W_A = pW_{AA} + qW_{Aa}$; $W_a = pW_{Aa} + qW_{aa}$. These can be used to predict which allele will increase in frequency at a single time step, but they will usually be changing and they are mathematical constructs derived from the genotype fitnesses. It has been argued that a particularly important case is that of *heterozygote superiority*, where W_{Aa} is higher than the other two. Then being an *A* allele is a good thing if your partner is an *a* allele, not so good if your partner is another *A*; no allele has a context-independent advantage over the other. Suppose that *Aa* is fittest and the other genotypes have equal fitness. Then the population will move to a stable equilibrium where $p = 0.5$. Each allele is favored when rare. When $p = 0.5$ the average fitnesses of the alleles are equal.

It is possible, however, to rewrite genetic models like this in a way that does not assign fitnesses to the diploid combinations (W_{AA}, W_{Aa}, etc.) at all, but assigns fitnesses to alleles in a way sensitive to the way alleles appear in different "environments" (Kerr and Godfrey-Smith 2002). Here a relevant part of an allele's environment is the allele at the same locus on the other homologous chromosome. It can be useful to "gestalt-switch" between two ways of looking at such a model, treating an allele's companions in the genome as its context or environment some of the time, and treating all the genes as parts of a larger biological unit or collective at other times.

Once we are working within a model in which there are particle-like genes, explanations of success in terms of context-sensitive fitnesses of the particles themselves are fine. But those particle-like genes do not correspond very closely to what genetic systems contain. Here there has been a historical shift. When the classic genetic models of evolution were developed, especially in the 1930s and the next few decades, the particle-like genes in those models were thought to correspond fairly well to the lower-level facts about genetic systems. One of the architects of these models, R. A. Fisher, argued in 1930 that inheritance *had* to operate in a "particulate" manner, with discrete and stable genes, in order for sustained Darwinian evolution to be possible. Fisher was writing before the discovery of the structure and role of DNA. The contrast he had in mind was with views in which inheritance involved a literal blending of material from the two parents, like mixing paint. This would lead to variation, the fuel for evolution, being quickly lost.

Genetic inheritance did turn out to be "particulate," but the particles are the single nucleotides. A single nucleotide can be seen as a gene in special cases, but single nucleotides do not play most of the roles that genes play in evolution or development—there are only four of them, for instance—and there is no larger unit that is "particulate" in the way Fisher supposed. The newer picture of genetic systems is one in which the match between a model of evolution as a competition between alleles and what really happens in genetic systems is more partial. The point is not merely that genes are more indefinite and blurry entities than had been supposed; it has to do with *why* they are less particle-like. Genomes, at least in organisms like us, are more organized entities, with large proportions of an organism's DNA engaged in subtle processes of regulation of the expression of "coding" regions. New genomes are made by combining large chunks of this genetic material from the genomes of each parent, and this is not much like shuffling a collection of alleles and stringing some together in a line.

Before summing up I will put on the table another set of phenomena that any view in this area has to deal with. Sex, as we saw, enables one piece of a genome to be passed on while other pieces are not, and hence for small pieces of genetic material to proliferate because they have good effects across many backgrounds. Once there is a way for one piece of a genome to be passed on when other parts of the same genome are not, there is the possibility of stretches of DNA doing things that give them an advantage over other bits of genetic material *within* the same organism (Burt and Trivers 2006). One example is *segregation distorter* alleles, which subvert the usual "fairness" of meiosis. When sex cells are made in an organism with two different alleles at a locus—an organism represented as *Aa*—usually half the sex cells carry *A* and half carry *a*, but a segregation distorter will get itself into more than half of the organism's gametes, by sabotaging production of gametes containing the other allele. Another example is a certain kind of "jumping gene," a gene that can move around within the genome. A *LINE transposon* codes for an mRNA molecule which is translated to produce a couple of proteins that bind to the mRNA and reverse-transcribe the RNA back into the cell's genome in a new location. So there are now two copies of that element in the genome where before there was one. I see these phenomena not as indicative of the general nature of gene action, but as derivative cases—no more a model of how genes generally operate than parasitism is a model for all of life.

In sum, DNA is both memory and part of the control system of cells. Because DNA is copied, it can undergo evolution in its own right. In a sexual population a stretch of DNA can persist in the form of copies for vast periods of time, even though each organism's genome comes and goes. Classic models of evolution seize on this fact and make it the basis for a representation of long-term change; evolution becomes change in the frequency of alleles in a gene pool. In the second chapter I distinguished *abstractions* from *idealizations* in science. Abstraction involves leaving things out, idealization involves imagining things to be simpler than they really are. Some debates about genetic models of evolution can be expressed by asking to what extent classical models of natural selection on particle-like genes are not merely abstract, but also engage in idealization—not only in obvious ways, in their treatment of mating systems and the like, but in their treatment of genes themselves.

Certainly these models track just a few features of a complex set of processes. And objects that look indefinite and vague up close can become usably sharp once you are looking from further away. Just about *any* object that looks sharp at the macroscopic level has blurred boundaries, and continuity with its environment, if you look close enough. A human being has unclear boundaries when we think about the role of bacterial symbionts, but that does not matter once we are doing demography or economics; then humans snap into focus as units. I criticized Richard Dawkins earlier when he said that genes have a "flintlike integrity." But even *flint* does not have a flintlike integrity if you look at it from sufficiently close up. Similarly, when looking at genetic systems in a fine-grained way, the gene has been de-particlized. But evolutionary biology works at a different scale. Perhaps when we look at change over a long period in an entire species, genes come into focus and evolution does look like change in allele frequencies.

That is one perspective. The other, which I am more sympathetic toward, sees our understanding of genetic systems as more at odds with a particle-like model of gene competition. Treating genes this way should be seen as a deliberate simplification. The point does not merely concern a blurring of boundaries, but a different picture of the genome at all scales. Genomes are more organized objects, and their partition into genes more artificial, than the classic models suppose. Here I put the opposition between two views in a stark form. Mixed and intermediate options are possible, and it is also likely that the next 50 years, perhaps the next 10, will change our view of genes and genomes once again.

FURTHER READING

On genetics, Falk (2009), Griffiths and Stotz (2013); on genes, Burian (2004), Gerstein et al. (2007); on causation, Beebee et al. (2010); on genes and environment, Oyama et al. (2001); on information, Oyama (1985), Kay (2000), Maynard Smith (2000) and see note 4; on genes and evolution, Sterelny and Kitcher (1989), Jablonka and Lamb (2005), Haig (1997, 2012).

CHAPTER 7

Species and the Tree of Life

RECOGNIZING "KINDS" OF some sort is ubiquitous, perhaps inevitable, in thought and description. In the case of living things, *species* have long seemed particularly important; a species seems to be the basic *kind of organism* that something is. Some philosophical problems with species come from general questions about what it is to find—or invent—kinds and categories in nature. Others come from the meeting between some intuitive ways of thinking about species and the view of the living world we get from evolutionary biology.

7.1. FROM TYPOLOGICAL TO PHYLOGENETIC VIEWS OF SPECIES

A complete view of species includes two parts. One is a view of the *grouping criteria* that place organisms into species. The other is a view of the *status* of species, a view of what sort of thing that group is. The same distinction can be applied to other kinds. (What groups organisms into the same genus, or phylum? What sort of thing *is* a phylum?) In this section the emphasis will be on grouping criteria.

A good place to start is with a *typological* view of species. On a typological view, organisms can be divided into types, where every individual of a type possesses an underlying nature, a set of distinctive internal properties, that is characteristic of that type and not of others. Views like this go back at least to Aristotle. Some psychologists think that a typological view of living things is something that humans also apply less consciously, as a result of habits that have a basis in our own evolutionary history.[1] Ty-

[1] See Medin and Atran (1999) and Griffiths (2002).

pological views are also often described as "essentialist," an idea I will discuss later.

Thinking about species was transformed by Darwin and evolutionary theory. The story is sometimes told as one in which everyone before Darwin was in the thrall of a typological view, which Darwin exploded. The real history is more complicated (Winsor 2006), but Darwin certainly changed the landscape. From then onward species had to be regarded as things that can evolve slowly from other species and have vague boundaries. On a Darwinian view, variation within a species does not reflect imperfection or faulty realization of a type, but is the normal state of affairs—I will say more about this attitude to variation at the end of the next chapter. It is possible to shoehorn evolutionary thinking into a typological view, but since Darwin, there has been a search for a treatment of species that fits better with an evolutionary perspective. At least two dozen different "species concepts" have been proposed, though they fall (appropriately) into a smaller number of clusters.

A simple way to move away from a typological view is to treat species as collections of organisms that are grouped by *overall similarity*, with respect to features like their shape, structure, and physiology. This view accepts that species contain variation and one species can shade into another, but holds that definite "clusters" of organisms can still be recognized. This is known as the *phenetic* view of species. Pheneticism is associated with a low-key attitude to the status of species and other taxonomic categories in biology; it sees these classifications as tools, and best constructed in advance of any theorizing about "real" units in nature.[2]

Some problems with this approach concern the very idea of "overall similarity," which depends on which properties are chosen and how they are combined in an overall measure. But suppose that overall similarity makes sense; there are also cases where it seems to give unwanted answers. First, in cases of "sibling species," for example in fruit flies, two or more species in a group are almost indistinguishable. On the other side, there are

---

[2]For a classic statement, see Sokal and Sneath (1963). For discussions of pheneticism and its aims, see Hull (1970) and Lewens (2012).

"polytypic" species, which include several "phenetic clusters" within what is usually seen as one species. So far it would be reasonable for a defender of the phenetic view to say that these arguments are question-begging: similarity is what counts, so there are not as many species of fruit fly after all. Perhaps, but the consequences of this reply would be dire. In some species, males and females are very different. Males may be tiny parasites attached to much larger females ("dwarf males" in barnacles, angler fish, and argonauts). If someone really thought about species in terms of overall similarity, they would treat the males and females as different species.

In the case of sex differences it is breeding that seems to tie different-looking organisms into a species. And perhaps this is all we need for an analysis. Perhaps a species is a *reproductive community*—a collection of organisms who can breed with each other and not with organisms outside that collection. The most famous version of this view was developed by Ernst Mayr (1942). According to Mayr's *biological species concept*, species are "groups of actually or potentially interbreeding natural populations which are reproductively isolated from other such groups." Calling this "the biological" species concept is misleading, as various other concepts of species are certainly biological. So I will generally call it the *reproductive community* analysis. This is a view in which species have a different status from other taxonomic categories, such as classes, phyla, and so on. Those "higher taxa" are often seen merely as collections of species that we have decided to group for reasons of convenience. Species, in contrast, are real evolutionary units.

Mayr, in the quote above, defined species as groups of *populations* rather than groups of *organisms*. How do we know which "population" an organism is a member of? The idea we want to capture is that species are collections of organisms who can breed with each other. But even setting aside sterile individuals, not all pairs of organisms within a species can breed together; two individuals of the same sex cannot. So we might say something like this: a species is a collection of organisms who all have the capacity to share *descendants*. These might be shared grand-offspring rather than shared offspring.

The version of the reproductive community view quoted above said that organisms in a species are those that "actually or potentially" interbreed. This seems to be a compromise between two thoughts. You might say that a species is a collection of organisms connected by *actual* reproductive relationships and distinct from other such collections. That, however, seems to make a lot depend on matters of accident. So it is tempting to add the idea of "potential" breeding. This leads to other problems. Suppose two groups of organisms do not in fact interbreed, even though they live in the same environment, because of a preference for their own kind during mate choice; there is no physiological impediment to their breeding together. Various kinds of "Hamlet fish" (*Hypoplectrus*), for example, who live in the same waters *can* breed together, but prefer not to. Is this one species or several? Some other groups of fish, in the "cichlid" group in Africa, show similar preferences, except that increasing murkiness in their water means that they can no longer reliably tell their preferred kind from others, and interbreeding is becoming more common.[3] Is this a case where there was always one species because there was "potential" interbreeding, or a case where a species barrier is falling? Biologists tend to say the latter. Mayr himself eventually dropped the term "potentially" from his definitions of species (1969). There are also cases with "chains" of reproductive compatibility, where (in a simplified version) A can breed with B, B can breed with C, but C can't breed with A. A famous case involves populations of salamanders that form a ring around the Central Valley in California. In a "ring species" situation of this kind, do we have one species or several?

A larger problem is that bacteria, and other asexual organisms, do not form reproductive communities in the relevant sense at all. It is not that bacteria never exchange genes. They do, but in a more haphazard way than organisms like us, and they may swap genes with distantly related groups (Franklin 2007). Bacteria do not form reproductive communities, but they do seem to form species. At least, there *seems* to be a lot of apparently meaningful and useful talk about bacterial species such as *E. coli*.

[3] For these cases see Ridley (2007, chap. 13).

A rival to the reproductive community view that deals better with asexual organisms is the "cohesion" analysis of species (Templeton 1989). This view holds that a species is a group of organisms unified by *some* factor from a range of "cohesion mechanisms." Sex is one such mechanism, but others are ecological; they involve living similar lives in a similar environment. Van Valen (1976) and others have developed views of species based entirely on these ecological factors.

I'll now introduce a problem with *all* the views above. They work fairly well when sorting into species the organisms alive at a single time, but have serious problems when we compare organisms alive at different times. This is because unless evolution is abrupt, within any lineage there will be gradual change in reproductive compatibilities, gradual change in phenotype, gradual change in lifestyle and ecology. The point of the idea of species is that it *partitions* living things into those that are in species S_1, those that are in species S_2, and so on. A partition is a set of categories that do not overlap, so that each object is in one category or the other, and every object is placed in some category. "Partition-thinking" mostly works fairly well (if we allow for exceptions and borderline cases) when comparing organisms present at a time, but it is much less suited to description of change over time.

An option that handles time better is the *phylogenetic* species concept. This approach starts from the idea that life on earth forms the shape of a "tree," an idea discussed later in this chapter. The tree of life represents the total set of ancestor-descendant relationships between living things on earth, and its shape (for at least many organisms) is a series of branchings, in which one lineage splits into two. If the tree is a real feature of life on earth, then maybe we can just say that *species are the twigs*. More precisely, species are segments of the tree between branching points (Cracraft 1983). This view derives from Willi Hennig (1966), the founder of the "cladist" approach to biological classification. It is a relative of the reproductive community view, but where the reproductive community view asked "who can you breed with?" the phylogenetic view asks "who are your ancestors and descendants?"

The phylogenetic approach also has other advantages. How do we tell the species of a sterile worker bee? Not by looking at who it can breed *with*, but who it was bred *from*. This has always been a lurking problem for the reproductive community view, a problem evaded by talking about interbreeding "populations"; as I said above, we then need to be told how to group organisms into populations. Phylogenetic views also have consequences that are surprising, though. There are different versions of the phylogenetic view and some of the oddities differ across versions. On one version (seen in Hennig himself), a species goes extinct whenever a new species is formed from it by a branching event; with each branching, one species goes extinct and two new ones form, even if one of the "new" ones is indistinguishable from the "extinct" one. This conclusion can be avoided by saying that if one of the new branches is much larger and the other is a small "budding," something that probably often happens in nature, then the old species has lived on in the larger branch. Another consequence— and one seen in all versions of the view—is that a species persists as long as there is no branching event, no matter how much change occurs on that lineage.

Phylogenetic views also have problems with asexual organisms like bacteria, not because there is no branching of lineages, but because there is too much; every event of reproduction is a branching. When thinking about bacteria, something like a phenetic view looks attractive again. Microbiologists often apply views of species that are at least partly phenetic (Ereshefsky 2010). And perhaps things are not so bleak for that approach after all, even outside of bacteria. The original versions of the phenetic view appeared and declined in the middle of the 20th century. Since that time, the rise of molecular biology has brought enormous changes to how people measure similarity between organisms, and this has led to a partial revival of the phenetic approach. Why not forget the idea of "overall" similarity and focus on *genetic* similarity?

Occasionally people opposing typological views claim that there are *no* genetic features unique to each species.[4] If the claim being made is that it would not be possible to determine from

---

[4]See, for example, Okasha (2002, p. 196): "It simply is not true that there is some common genetic property which all members of a given species share, and

its genome whether an organism alive now is (say) a member of *Homo sapiens* or of some other species, then the claim is probably false. Species are genetically variable, but that is compatible with their having diagnostic features. Thinking within a phylogenetic framework, in which species are parts of a tree-like structure, we would expect each lineage to accumulate genetic peculiarities, showing marks of its place in the tree of life. The genetic profile characteristic of a species need not—probably will not—take the form of stretch of DNA that all members share in exactly the same form, but this is not necessary for a genetic profile to be real and recognizable. A "genetic cluster" analysis of species can be applied to both sexual and asexual organisms (Mallet 1995).

Let's assume that all or most species do have genetic profiles that enable us to sort the organisms that exist at a time into one species or another. That does not settle all the questions, and I think that writers denying that species are genetically distinctive may have had other issues in mind. There is no reason to think that a genetic profile characteristic of a species at one time will stay fixed within a lineage. And suppose we identify a set of distinctive genetic features of (say) our own species, *Homo sapiens*. Would it be *impossible* for an organism to live a recognizably human life without these genetic features? As far as I can tell, that is a question we don't know the answer to (even being relaxed about the ambiguities surrounding the idea of impossibility). It would not be surprising if there were some human-specific genes that were also indispensable to living a human-like life, but it would also not be that surprising if there were not; the loss of these genes might be compensated for by the presence of others. And once we leave very distinctive species, like ours, and consider cases where there are large numbers of different species (of fish, of fruit flies) that are very similar in how they live, it becomes clear that the lifestyle of one species can be lived with the genome of another. Humans, who have no close living relatives and who live such unusual lives, are not a good model for species in general.

which all members of other species lack." See Devitt (2008) for discussion of these claims.

At this point it might seem reasonable to say we should not be trying to choose between these species concepts, many of which make good sense but employ very different criteria. Perhaps a "pluralist" attitude is the right move; people should make use of different species concepts for different purposes.[5] Maybe. But a different reaction is also possible. Perhaps if there is not a single grouping criterion applicable to all cases, this is a problem for the whole idea of a biological species. Perhaps talk of "pluralism" is a way of refusing to accept that the concept of a species has *disintegrated*—it has not just divided, but collapsed.[6]

A related position is that "species-talk" can be useful in biology even though species are not real units in the natural world. I've become more and more sympathetic to this view through the writing of this chapter. This may be due to the constant focus on puzzle cases, but from an evolutionary point of view, the species concept can come to seem like an attempt to partition the unpartitionable. When thinking about this view, the problem of categorizing organisms living at different times is the one to focus on. Ernst Mayr said that his biological species concept was intended to be applied in a "nondimensional" way. This term is obscure, but he meant that this species concept should not be used to compare organisms alive at different times. Mayr said that as two organisms get further apart in time, it gets less and less important to talk about whether they are in the same species. That may be right, but if species are real units in nature, then for all pairs of organisms, no matter when they were alive, there ought to be a fact of the matter about whether they are in the same species or not. This principle should not be applied with obsessive strictness; there might be borderline cases and exceptions. But to say that the idea of a species is supposed to be applied only to organisms present at the same time, and that for most pairs of organisms the question "are they in the same species?" should not be asked, is to give up on species as real units in nature.

---

[5] See Kitcher (1984), Dupré (1999).
[6] For views of this kind, see Ereshefsky (1992, 1998) and Mishler (1999).

7.2. Particulars, properties, and kinds

Earlier I distinguished views about *grouping criteria* from views about the *status* of species. We left the grouping criteria in a state of disarray. This section will look at the other issues. Questions about the status of species plunge us into metaphysics, the part of philosophy concerned with the most general questions about the nature of reality.

When someone asks what sort of thing a species is, the natural first answer is that a species is a *kind*. What is a "kind"? Apparently, a kind is a collection of things that are unified by having some property, or properties, in common. Michael Ghiselin (1974) and David Hull (1976) argued that this is entirely the wrong road to go down when thinking about species. A species is an *individual*.

I used the term "individual" a lot in chapter 5. The Ghiselin-Hull view about species uses the term in a broader sense. Roughly speaking, what they were saying is that a species is a *particular* thing, a single object with a location in space and time. Their main argument is that species must be individuals if they are to play the role that evolutionary biology attributes to them. Species evolve. They come into being, change, and disappear. Those are things that only individuals can do. So although it is familiar to talk about organisms as *instances* of their species, really we should say that an organism is a *part* of its species—it is one physical part of a large scattered object. The relation between you and the human species is the same, logically speaking, as the relation between your left hand and your whole body.

I will approach this issue by first making some general distinctions that involve ways of grouping things. One way to think about groupings is with the idea of a *set*. A set is a collection of objects, the *members* of the set. Sets are often seen as "abstract," like numbers and mathematical functions, even when the set's members are physical objects. A second way of thinking about groupings is through the idea of a *sum*. A sum (in the sense relevant here) has parts rather than members. Your left hand is a part of your body, not a member or instance of your body. The sum of your body parts has a weight and a location; it can be moved about. Many philosophers would deny that sets can be moved about.

A third way of thinking about groupings is with the idea of a *property*, and the idea of the *sharing* of properties. Various tables share the property of being made of wood. There are hints of metaphor in this language of "sharing," and the idea can be expressed in more technical language, which may or may not make things clearer: the tables all *instantiate* the same property.

How does the idea of a "kind" fit in? I see it as lurking between the idea of a sharable property and the idea of a set. Sometimes talk of kinds is talk about the objects making up the kind. Then a kind looks like a set. Sometimes talk of kinds seems to be talk about whatever it is that groups the objects that make up the kind; this is closer to a property. As the word "kind" is ambiguous in this way, I will focus on the three things that do look distinct: sets, sums, and properties.

How should we think about species? Some philosophers have given arguments against the reality of sets, or of properties, even occasionally of sums, that are intended to be entirely general. If sets do not exist at all, then if species are real they cannot be sets. I won't try to sort out those issues here. I'll assume that all three ways of thinking are acceptable in principle. The question is how they relate to species. My suggestion is that all three ways of thinking about species are OK. Species are not *really* sums (big scattered particulars), or really sets, or kinds defined by shared properties. Species are aspects of the world's organization that can be thought about in three different ways. You might wonder how we could know this when we have not decided on a grouping criterion for species. But roughly speaking, this message applies for all the grouping criteria discussed earlier, or at least all the views that treat species as real.

I will make this argument with the aid of two extra examples. One example is *carbon dioxide*. The other is *the Churchills*. When I say "the Churchills" I mean the collection of people who made up Winston Churchill's immediate family. (There is nothing special about the choice of Churchill—you can pick any family you like.) What is the "grouping criterion" for the Churchills? It is a matter of parent-offspring relations, and other causal relations such as marriage. Once we know that someone is a human being, to be a Churchill is not a matter of similarity to other Churchills; it is a matter of being causally connected to Winston and the other Churchills in the right way. This is reminiscent of a phylogenetic species concept. But as far as metaphysics goes, all three ways of thinking about the Churchills are possible. You can think of them as a set, with individual people as members. You can think of the family as a sum, an object with parts. And you can think of the property of *being a Churchill*, which they all share. The three modes of grouping enable us to say different things. To say "The Churchill family left London" is to treat the family as a particular, as something that can move through space. If you say "All the Churchills liked cigars" you are probably thinking of them as a set. But you can also say "It was being a Churchill that got him that job." Then you are using the language of properties.

Unlike the Churchills, *carbon dioxide* is a kind that can be understood in a "typological" manner. To be a molecule of CO_2 is to have one carbon atom and two oxygen atoms arranged in a particular way. In this respect, being carbon dioxide is nothing like being a Churchill. But all three ways of thinking about groupings are available here, too. You can say that dry ice is made of carbon dioxide (property); you can say that all carbon dioxide molecules contain 22 protons (set); and you can say that the CO_2 in the earth's atmosphere has increased greatly during industrial times (sum).

All three ways of thinking about carbon dioxide and the Churchills are available, and the same applies to chimps—to the species *Pan troglodytes*. Some people think of the chimp category as similar to the Churchill category—defined by ancestry. Others think of it as more like carbon dioxide. *Either way*, you can treat it in terms of sets, sums, and properties. You can see the species as a large particular—something that came into existence at a particular time, part of the totality of life on earth. You can also see the chimp species as a set of organisms, and you can talk of *being a chimp*, a feature shared by various objects, and hence also think of chimps as a *kind* of thing.

Talk of sets, sums, and properties are three ways that we humans have of picking out structure in the world. These ways of grouping things can be called "ontological frameworks" (or just "frameworks"). Though all three are *available* in all the cases discussed so far, some frameworks fit more naturally with some grouping criteria than others, and some grouping criteria are more useful than others in specific cases. The language of properties is especially useful when dealing with things like carbon dioxide, and the language of some is more useful when dealing with things like the Churchills. *Being a Churchill* is an odd-looking property, and the scattered object made up of all the world's CO_2 is an odd-looking particular (so much so that you may have questioned my example at the end of the paragraph two prior to this one).

What is the difference? One difference can be expressed using another metaphysical distinction, between *intrinsic* and *extrinsic* properties of an object. Intrinsic properties are properties an object has that do not depend on the existence and arrangement of other objects. An object's chemical composition is intrinsic. Extrinsic properties do depend on what other objects there are and what they are like.[7] Whether something is a molecule of CO_2 is intrinsic; whether some person is a Churchill is extrinsic. Extrinsic properties are often called "relational." Some special cases make the term "extrinsic" better, but extrinsic properties are indeed a matter of relations. Extrinsic properties have been regarded by many philosophers as second-rate, or less real, than intrinsic properties. When a view like that is used to treat relations themselves as somehow second-rate, I think it is mistaken. But the *language* of "properties" does apply most naturally when properties are intrinsic. When relations are very important, it often becomes natural to talk of parts and wholes. Consider the components of your body: you can talk about a set of parts that make up your body, and you can talk of the property *being a part of my body*. But part-whole thinking seems more useful here most of the time. This is because what unifies those parts into a single body is something about the relations between them, not their intrinsic features, and the operation of your body is very sensitive to the relations between the parts that make it up.

Returning to species: phylogenetic and ecological properties of organisms are extrinsic. Genetic and "phenetic" properties are intrinsic. (The properties used in reproductive community analyses

[7]See Langton and Lewis (1998) and Weatherson (2006) for detailed treatments.

of species are sometimes called extrinsic, but these are a more complicated case.) If you think of species phylogenetically, it will be natural to think of them as particulars. If you think of species genetically, it will be more natural to think of them as kinds unified by shared properties. Further scientific ideas might also push you one way or the other. Some defenses of the species-as-individuals view emphasized the internal "cohesion" of species, including a tendency that species have to maintain themselves over time. To the extent that species do things like this, a species is more like a human body than it is like a collection of CO_2 molecules floating about in a gas. Those claims about the "cohesion" of species are controversial, and some argue that smaller units than species—local populations—are more important in evolutionary explanations anyway (Ehrlich and Raven 1969). If this last claim is true, there will be less motivation to speak of species as large particulars, though it will still be possible.

What about sets? Talk of sets brings less with it than talk of sums and properties. Set-talk is both less constrained and less suggestive. Partly for this reason, sets are popular in philosophy as a general-purpose way of treating groupings. Many kinds of structure can be modeled within the framework of set theory. In the debates about species, Philip Kitcher (1984) took this option: a species is a set, he said, and both intrinsic and extrinsic properties might matter in determining which species an organism is in. Kitcher is also a pluralist about species groupings.

A connection here can also be made to another controversial concept, the idea of an *essence*. This term is often associated with a typological conception of species—"essentialist" and "typological" are sometimes used interchangeably. An essence is an internal property that makes something what it is. However, there are philosophers who accept reproductive community analyses, or phylogenetic analyses, who want to say that these are not rejections of the idea of a species essence, but different views of what the essences of species *are*. Perhaps the essence of a species is its position in the tree of life.[8]

---

[8] See Okasha (2002), Griffiths (1999). For arguments related to my own here, see Ereshefsky (2010).

It is *possible* to talk of something's "essence" being extrinsic, but the idea of an essence applies more naturally to intrinsic properties. Historically, essences have been seen as properties "expressed" in the behaviors and observable features of the objects that have them. This idea of "expression" is certainly suspicious, but at least it seems right to say that the internal nature of something can be *manifested* in how it behaves. This makes sense in the case of CO_2. But extrinsic properties, on their own, cannot give rise to the behaviors of an object in this way. Though they can be causally important, can they be manifested or expressed? To use the word "essence" for extrinsic properties is to hang onto a loaded word when much of its original meaning has been stripped out.

Different kinds of structure in the world make different grouping criteria useful, and different grouping criteria often work naturally with one ontological framework or another. Ghiselin and Hull were right that there is something important in the idea that a species can be seen as a large particular with organisms as parts. This was a valuable shift of perspective, even if it is not true that species are individuals *rather than* sets.

A last comparison that can be made here is with the Internet. "The Internet" is a name for a large object, a particular, scattered through space, with many individual computers, with their files, and other machines as parts. In 1980, there would not have been much reason to talk about the collection of computers in the world as a sum, a big individual; there was just a set of computers. But the pattern of interaction that exists between computers and their users now makes it natural to talk about a large object, the Internet, which continues to exist even when many of its constituent computers are not interacting with each other.

7.3. THE TREE OF LIFE AND THE ORIGIN OF SPECIES

In the first chapter I distinguished two parts of Darwin's theory of evolution: the idea of change by natural selection and the idea of a "tree of life." The tree of life has become one of the most important organizing ideas in biology. It is an immensely suggestive idea, but what exactly is the tree supposed to *be*? Sometimes the tree of life is described as a "metaphor," at other times as a "discovery." But if the tree is a discovery, it should not be a metaphor—or at least, it should be possible to restate the "discovery" part in non-metaphorical terms. Most of this section will discuss the tree of life in its own right, but I also want to use the tree to help us with species, and we'll come back to them at the end.

As noted in chapter 1, people had been thinking about life in tree-based ways for some time before Darwin. Darwin's innovation was to think of the tree in genealogical terms, as representing relations of ancestry and descent. Often these "descent" relations are seen as relations connecting species themselves, but a good way to approach the tree is to start by making no assumptions about species, and think just about individual organisms.

Imagine taking a huge transparent cube of jelly and drawing a mark inside it for every organism that has ever lived, starting from the beginning of life at the bottom of the cube and working up. The vertical location of the mark represents when that organism was alive. Its exact location in the two other dimensions does not matter, except that offspring are always drawn somewhere just above their parents and moving away from the center because more and more organisms have to be fitted in as you get higher. You also draw lines between the marks to indicate parent-offspring relationships. The process is continued to include every organism that has lived on earth. What does the result look like? From up close, you would see an intricate network of lines, splitting and joining. But in at least some regions, you would also see that these lines tend to collect together into fairly definite threads and strands. If you could somehow grab a few strands and pull sideways, you could find that most of the network does not come along with the ones you are pulling. If you pulled all the strands out sideways and then stepped back far enough for all the points and lines for individuals to blur into each other, what you would then see is the overall form of the genealogical relationships between different kinds of life on earth. This structure has different shapes in different regions, as I will discuss below. But in the region where organisms like us are found, what you would find has the shape of a tree, a succession of branchings always pointing up. Even where the shape is not tree-like, there are features that are common across the whole structure. At the top are the points for us and the other organisms alive now. If we imagine zooming back in to that vantage point, and looking downward, we would find a maze of connections stretching into the past, but one with an overall pattern of convergence. Any two points at the top of tree—any two organisms living now—can be traced to a common ancestor somewhere lower down.

In this description I assumed that life is, in fact, historically unified on earth. It might turn out that early on there were several separate originations of life, most of them dying out, which would lead to several unconnected shapes being drawn in the block. A different simplifying assumption is that it makes sense to count and represent "organisms" at the early stages of life on earth. That is probably a significant idealization, but I'll leave it in place.

Although the shape we've imagined drawing was made by means of many abstractions, metaphors, and conventions of drawing, the result would be a representation of some real facts about life on earth. What has been drawn is a kind of *map*, like a subway map of a city, something whose structure faithfully represents one kind of connectedness that exists in the terrain being mapped.

Thinking about the tree of life is a further application of the style of thinking that Hull and Ghiselin employed with their "species as individuals" claim, discussed in the previous section. The gestalt-switch into thinking of all life as a connected object is vividly expressed also in this quote from Julian Huxley, writing in 1912:

As individual emerges from individual along the line of species, so does species emerge from species along the line of life, and every animal and plant, in spite of its separateness and individuality, is only a part of the single, continuous, advancing flow of protoplasm that is invading and subduing the passive but stubborn stuff of the inorganic. (J. Huxley 1912, pp. 27–28)

When thinking of life on earth as a single object in this way—perhaps in less imperialistic terms than Huxley's—we find that life on earth has a physical *shape*, a shape in space and time, which can be discovered by biology. The *total* shape of life on earth, the sort of thing Huxley is thinking of, includes facts about where each organism lives, who eats who, and so on. The shape imagined earlier in the block of jelly ignores most of this. But the tree shape in the block is a representation of the *distinctive* features of the shape of life hypothesized by Darwinism and modern evolutionary theory. Any theory that recognizes the ongoing "spontaneous generation" of life, such as the theory of Lamarck, differs from Darwinism about aspects of the shape of life represented in the tree. Worms for Lamarck were like CO_2 for us; they come into existence when circumstances are right, without an ancestral connection to other worms. In some discussions the tree of life is seen mostly as a tool for explaining some particular facts about organisms and their similarities. I agree that it has these uses, but the tree has a role that is more general. It is an attempt to represent some facts about the shape of life on earth.

The way I introduced the tree was by starting with relations between organisms, so any pattern of common ancestry that links species (whatever species might be) has its basis in a pattern of common ancestry linking individual organisms. Let's take a closer look at the relationships between zoomed-in and zoomed-out perspectives. One of the first people to think carefully about the tree of life in this way was Willi Hennig, whose book *Phylogenetic Systematics* (1966) revolutionized classification in biology. Hennig argued that all talk about species and other such categories is a coarse-grained depiction of a total "fabric" of relationships between individual organisms. A modified version of a diagram used by Hennig to represent the relations between different scales was used back at the start of chapter 3.

In Hennig's diagram, which is a microcosm of the tree of life as I described it earlier in this section, assumptions are made that make the drawing operation easy. The parent-offspring relations drawn between individual organisms are sexual—two arrows go up to each new individual. As discussed in chapter 5, much of life does not behave like this. Many organisms are asexual, either always or some of the time. Organisms like ferns have an "alternation of generations" between sexual and asexual stages. In plants like aspen, it is hard to distinguish growth from reproduction. Suppose Hennig tried to draw his figure for bacteria. Here the parent-offspring relationships between organisms are branchings, with one giving rise to two, and genes are also swapped and passed around in "lateral gene transfer." So there is a low-level branching structure generated by cell division, plus occasional "bridges" between lineages through lateral gene transfer, with some bridges connecting distantly related organisms. The tree-like shape also dissolves in other ways when we are dealing with unicellular life. Perhaps the most important features that are not tree-like are the great symbiotic marriages that produced the eukaryotic cell, especially the process that produced our mitochondria (§5.2).

So at different places on the "tree" of life (suddenly I am using scare quotes), we find different organism-level relationships when we are zoomed in, and these have consequences for what shape can be claimed to exist when we zoom out. In the lower left of Figure 7.1 I have bacteria, with tree-like branchings at the organism level and occasional lateral gene transfer events (the dotted line). Solid lines are drawn for the zoomed-out prokaryotic lineages, even though that is controversial. At the lower right I have zoomed in on the symbiotic event that produced mitochondria,

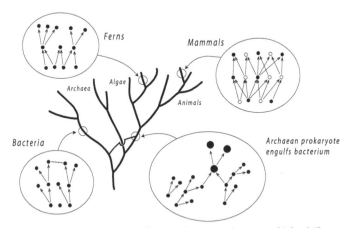

Figure 7.1. Zooming in to different places on the tree of life, different lower-level relationships are found.

a cross-link in the tree. These parts of the diagram zoom in to an "organism" level, where both bacteria and mammals are counted as organisms. But mammals are, of course, collections of cells. We could zoom in all the way to a tree of cells, and for some biologists, the tree of cells is *the* tree of life. It is also possible to construct trees for individual genes.

Some biologists think that people have hung onto "tree" talk for too long, given how many exceptions are known, and we should start talking of the "net" of life. This does not affect the idea that a total representation of parent-offspring relationships between individuals, like the one I started this section with, is a map of real features of the shape of life on earth. It only concerns what the map looks like. In the future the "tree of life" may be seen as a rough representation that has been superseded by something else, which might be just more diverse in shape or might have other theoretically important structure.

To finish this section I will link this discussion of the tree to the debates about species. What we have found is that at different places on the tree (or net) of life, there are not just different organisms, but different *kinds of kinds*. We find different ways in which organisms relate to and resemble each other (Dupré 2006). These differences are a consequence of the mode of reproduction, the behavioral tendencies, the genetics, and other features of the organisms in the kind. In some parts of the tree we find collections of organisms bound together by sex, forming a "fabric-like" reproductive community that is sharply distinct from other reproductive communities, at least at a time. In other parts of the tree we find sex but less cohesive reproductive communities. This is the case in plants, especially, where occasional and not-so-occasional hybridization occurs in many groups. In yet other parts of the tree, gene exchange is rarer and haphazard, so the clustering of organisms into recognizable groups—to the extent that we find it—has a different explanation. We find different ways that living things are bound together into groups, and into other structures, as a consequence of the biology of the organisms themselves.

If that is how the biological world looks, where does this leave us with the idea of *species*? Various different ways of using the term make sense, though all are somewhat at odds with its previous meanings and associations. One possibility is to use the term broadly, for many different kinds of kinds. Then "species" will refer to different sorts of units in different parts of the tree, but it will make sense to ask questions like: how do the patterns of diversity within microbial species relate to those in animal species? Another possibility is to treat a *species* as one kind of kind, and an evolutionary outcome with a history that is tied up with the history of sex. There is no point in trying to legislate about use of the word. A term surrounded by a long history of debate and diverse applications, as "species" is, will take its own undirected evolutionary path.

FURTHER READING

On classification and kinds, Dupré (1993, 2006), Hull (1988), Boyd (1999); on species, de Queiroz (1999), Wilson (1999), Coyne and Orr (2004); on essentialism, Wilson et al. (2007); on the tree of life, Dawkins (2004), Doolittle and Bapteste (2007), Franklin-Hall (2010), Velasco (2012).

Evolution and Social Behavior

THIS CHAPTER IS about social behavior, especially cooperation, altruism, and their relatives. These behaviors have great importance in human life, and they also pose problems for evolutionary explanation. If evolution is a reproductive competition, how can organisms evolve a tendency to give resources away, or to make sacrifices for others?

As theories have developed, it has become apparent that these behaviors play a multifaceted role: first as aspects of the social life of animals with complex behaviors like ourselves, second as features of interactions *within* organisms—between cells and between genes, for example—and also in symbiotic associations that make the boundaries of organisms unclear. The evolution of cooperation has become integral to our understanding of the workings and origins of living things.

8.1. COOPERATION AND ALTRUISM

This section surveys evolutionary explanations of cooperation and altruism. The topic here is *behaviors* that have positive effects on the fitness of other organisms; the psychological motives behind the behaviors are irrelevant, and the basic models apply to entities with no psychology at all. Psychology will be put back on the table in the next section.

Terminologies in this area are variable. In general I will talk about *cooperation* when two or more agents engage in an interaction that involves helping of some kind, perhaps including immediate sacrifices, but where everyone involved tends to benefit, either right away or in the long run. These cases can also be described as *mutualism*—mutualism because the interaction benefits both sides. The evolutionary problem here is explaining how these arrangements can be established given that there will often be opportunities to accept the benefits but avoid one's own contribution. They may also require subtle coordination of behaviors on each side. In the case of *altruistic* actions, a pattern of behavior evolves where some individuals *give away fitness*—they pay a net cost, in an evolutionarily relevant sense, in a way measured over their whole lifetime. Here the problem is to explain how such a behavior could ever survive. The border between the two is not always clear, though, and I will sometimes use terminologies that have become standard in particular discussions even though they do not conform to the distinction above. "Prosocial" will also be used as a general term for behaviors of these kinds.

The 20th century saw vigorous debate about three mechanisms for explaining the evolution of prosocial behavior: *group selection, kin selection*, and *reciprocity*. According to group selection hypotheses, altruism and related behaviors can survive because of benefits at the group level. Suppose a population is divided into groups, some containing many altruists who help each other, and some containing mostly selfish individuals who do not. The groups with many altruists may be more productive and resistant to extinction than the others. Darwin, in the *Descent of Man* (1871), thought that moral character in human tribes might be favored by selection acting on groups in this way.

For some years in the mid-20th century this form of explanation was overused (Williams 1966). This was the result of the acceptance in some parts of biology of a rather harmonious conception of nature. Harmonious groups, however, face problems of *subversion*. If a selfish mutant appears, accepting the benefits and paying none of the costs, it will do well. Once the selfish type spreads, it may drag down the group, but that will not stop it spreading. If a group *remains* altruistic, maybe it will survive and give rise to new groups, while selfish groups are going extinct. But that seems likely to be a slow process, while the subversion of groups from within seems likely to be a fast one. Since recognition of this problem, some have given up on the idea of group selection and others have refined it (Sober and Wilson 1998).

A second form of explanation for altruism was developed in the 1960s by William Hamilton (1964, 1975), explanations in terms of "inclusive fitness" or "kin selection." An organism can be altruistic as an individual but still behave in a way that benefits the genes it carries if its generosity is directed at its biological relatives—individuals who are likely to carry the same genes, including the gene for altruistic behavior itself. A third approach is based on *reciprocity* (Trivers 1971). An organism can gain a long-term benefit by making short-term sacrifices, so long as the favors are likely to be repaid. This third framework does not use the idea of benefits to groups, and does not require that the interacting individuals be related—they might even be from different species.

An influential framework for modeling these issues, especially suited to reciprocity but with more general applications, is *evolutionary game theory* (see also §2.3). The *Prisoner's Dilemma* has become a focal model for the problem of cooperation. The original scenario does not map very naturally to biological cases, but here it is. You and a fellow gang member are separately interrogated by the police. If you inform on your accomplice and he does not inform on you, you will be set free and he given a long sentence. If neither informs, there is a very short sentence for both of you. If he informs and you do not, you are the one who gets the long sentence. If both inform, you receive a short sentence, but longer than the one you'd get if neither of you had talked. You make your decision knowing that your accomplice has the same choice as you, faced simultaneously, and neither party can influence the other. This is a symmetrical two-player game. In the matrix below, the payoffs are those for the row player, "Player 1," conditional on what the other player does. "C" stands for cooperation and "D" for defect, where "cooperation" is with the other player, not the police.

		Player 2	
		C	D
	C	$R = 3$	$S = 1$
Player 1	D	$T = 4$	$P = 2$

We have a Prisoner's Dilemma as long as $T > R > P > S$. The Temptation to exploit a cooperator is better than the Reward for cooperation, which is better than the Punishment for mutual defection, which is better than being a Sucker who is exploited by his partner. I have given a numerical example that fits these criteria. (It is often also required that $2R > T + S$. Ignore the fact that in the original scenario, the outcomes were punishments rather than benefits; all that matters is the relations between the numbers.)

As far as rational choice goes, and assuming self-interest, your best option is to defect. If the other player cooperates, your best move is D. If he defects, your best move is still D. Defection *strictly dominates* cooperation: it is better no matter what the other player does. But the other player will think the same way. So both of you will defect, and both will be worse off than if you had somehow managed to both cooperate.

What if the two players tend to *repeat* the game? The *Iterated Prisoner's Dilemma* is a series of interactions ("trials"), each a Prisoner's Dilemma, between two agents. Each player can modify its behavior according to what has happened on earlier trials. Especially if it is uncertain how long the iteration will go on, it is not so obvious which choices are rational. This problem was the topic of a series of "tournaments" between computer programs organized by Robert Axelrod (1984). The aim was to work out which strategies would do well in a quasi-evolutionary competition. The most successful strategy overall was very simple: "tit for tat" (TFT, submitted to the tournament by Anatol Rapoport) cooperates on the first trial of its interaction with any partner, and then on subsequent trials it does whatever the other player did on the previous trial. TFT is initially cooperative, quick to retaliate, but quick to forgive. In Axelrod's tournaments it did better than much more complicated strategies, not by winning each head-to-head contest—it *never* does that—but by doing best on average.

TFT's success led Axelrod and Hamilton (1981) to suggest a scenario for the evolution of cooperation. Imagine an initial situation in which a population is full of individuals who defect no matter what. This is the "ALLD" strategy. TFT cannot get established in such an environment unless the first few TFT-ers interact mostly with each other. ALLD is *evolutionarily stable* against

TFT: as long as individuals are meeting at random, ALLD cannot be invaded when it is common.[1] But if TFT can get a foothold, and each pair's interactions are repeated for long enough, eventually a critical mass of TFT players is reached beyond which TFT does better than ALLD even if meetings in the population are random. Then TFT takes over and can resist reinvasion by ALLD and other exploitative strategies.

Subsequent work has shown that TFT is not as magical as it first appeared, but a fair bit of the general picture that Axelrod argued for has remained intact. In computer simulations of diverse collections of individuals who form pairs randomly, play Iterated Prisoner's Dilemmas, and reproduce according to their payoffs, giving rise to more individuals of the same kind, some degree of cooperation arises and becomes established fairly reliably, especially if the initial pool includes something similar to TFT (Binmore 1998, Nowak 2006a, Kuhn 2009).

It is tempting to treat these results as suggestive about practical issues as well as evolutionary ones. Axelrod saw the success of TFT as teaching him a moral lesson: the maintenance of a good social order requires that people retaliate. "Turning the other cheek" undermines cooperation, as it makes exploitation feasible. One can certainly see the point. But when thinking about TFT, I am reminded of a scene from the classic gangster film *The Godfather* (1972). Michael Corleone has left America and is hiding in Sicily. He is being shown around a village, and sees no men, only women, on the streets. When he asks "Where are all the men?" he is told, "They're dead from vendettas."

The debate about the relative importance of kin selection, group selection, and reciprocity has been intense. The mechanisms above make different assumptions: kin selection hypotheses suppose that specific genes are responsible for altruism and are being passed on by the relatives of individuals who make

---

[1] A strategy X is an *evolutionarily stable strategy* with respect to a given range of alternatives (Y, Z, . . .) if it cannot be invaded when nearly everyone in the population is playing it. Either X does better when interacting with itself than any alternative does when interacting with X, or if some alternative Y does *as* well when interacting with X as X does with itself, then X does better when interacting with Y than Y does with itself (Maynard Smith 1982).

sacrifices; reciprocity does not. Some group selection models assume an "altruism gene" (or some more complicated genetic basis), and some make use of cultural transmission of behaviors instead. Over recent years, though, a theoretical unification of these ideas has emerged. The main unifying idea was expressed by Hamilton in 1975, though it took a while for the message to become entirely clear.

Start with a model of *any* behavior where one entity's actions affect another. Assume two types, one (*A*) who performs an act that affects others and another (*B*) who does not perform it. Individuals might interact in pairs, or in larger groups or networks. The act performed by the *A* type has a direct effect on the fitness of the actor, and another effect on the fitness of everyone who that *A* interacts with. Both of these might be positive or negative. When there is a direct benefit for performing the act, there is no special problem of explaining why the *A* type does well. The most important case is explaining how an act that is harmful to self and beneficial to others can evolve.

Here as in any such case, what is needed is for average fitness to be higher for *A*s than for *B*s. How is this possible? The key is that *A*s are recipients as well as donors, and the outcome depends on who is interacting with whom. Assume first that interactions in the population are random—an *A* is neither more nor less likely than a *B* to interact with other *A* types. Then the benefits of the act become evolutionarily irrelevant, because they are equally likely to fall on anyone, and all that matters is the direct effect of the act performed by *A*. If the direct effect is a cost, the *A* type must have lower average fitness. But suppose the population is structured in such a way that *A*s tend to interact mostly with *A*s and *B*s with *B*s. Then the benefits of the *A* type's actions fall mostly on those who are themselves *A*s, and *A* can have higher average fitness. A mathematical treatment is given in Box 8.1.

The standard mechanisms for enabling the evolution of altruism are ways of getting this assortment of interactions to occur, ways of getting the benefits of a prosocial act to fall on those who are themselves prosocial. Acting altruistically toward relatives is clearly a case of this. A division of the population into groups who interact mostly internally does the trick too, though only if there are differences in the frequency of altruism across these groups. Groups themselves are not necessary for the assortment of types, as a population might form a network where individuals interact with their neighbors and there are no group boundaries. If the distribution of individuals on the network is nonrandom, the benefits of altruism can go to other altruists.

Another connection can be made by returning to the Iterated Prisoner's Dilemma and taking a closer look at TFT. Suppose there is a population with a fair number of TFT players and a fair number of ALLD. The players come together at random and interact for a reasonably large number of trials. Then although *strategies* meet at random, the presence of TFT generates positive assortment at the level of *behaviors* (Michod and Sanderson 1985). This is because of the role of copying in TFT. When TFT meets ALLD, the result is a long string of mutual Ds, with just one C-D combination at the start of the sequence. When TFT meets TFT, the result is a long string of C-C. When ALLD meets ALLD, the result is a string of D-D. So very few C behaviors are paired with D behaviors; most C behaviors meet other Cs, and the only exceptions are the first trials when a TFT meets an ALLD.

When TFT is rare and ALLD is common, this effect is weak. Almost every action in the population is a D and almost every C is paired with D. Some association at the level of strategies is needed for TFT to become initially established, as Axelrod and Hamilton noted. Once it has a toehold, TFT generates correlation at the level of behaviors and can flourish even if individuals meet at random.

Correlated interaction affects all sorts of issues in the area of rational choice and evolution (Skyrms 1996). Suppose we have a one-shot Prisoner's Dilemma being played in a population with random encounters. Defection strictly dominates cooperation; it is better no matter what the other player does. Defection can invade a cooperative population, and once it is common it cannot be invaded in turn. If interaction is correlated, things are different. Cooperation is still strictly dominated as far as rational choice goes, but cooperation can invade a population of defectors and can resist invasion once established. (This follows from arguments in Box 8.1). When interactions are correlated, even a strictly dominated strategy can evolve.

So far in this section I have focused on the Prisoner's Dilemma, the best-known game in this area, but I will introduce another. Suppose the numbers for T and R in the matrix on page 122 are switched. The result is a version of the *Stag Hunt* (Skyrms 2004), a game where $R > T \geq P > S$. Now the best response to a cooperator is to cooperate oneself. The best response to a defector is still to defect. There is no temptation to exploit a cooperator (so "T" is no longer an apt symbol for the payoff to a defector who encounters a cooperator). *If* you could have confidence that the other side would cooperate, you would want to cooperate too. The problem now is not exploitation; the problem is that *if*. There is no point in cooperating with a defector, and for each player, to defect is a safer choice overall; the worst outcome that a defector might receive is better than the worst outcome for a cooperator.

As the name "Stag Hunt" suggests, a relevant scenario here is cooperative hunting. Two agents can either cooperate and hunt big game, a stag, or go their own way and hunt hare. Much can be gained from working together, but it takes two to tango. Because prosocial interactions of this kind are *mutualistic*, benefitting both sides, the thing that needs explaining is not the viability of generosity or sacrifice, but achieving the kinds of coordination that give rise to a substantial payoff.

As evolutionary models of social behavior have developed, so have the breadth of their application. Fine-tuned cooperative relationships are important in many "evolutionary transitions" (Maynard Smith and Szathmáry 1995, discussed in §5.2). These transitions include the evolution of the eukaryotic cell from simpler cells, the evolution of multicellular organisms from single-celled organisms, and the evolution of cohesive colonies and societies. In all those cases preexisting biological units come together, cooperate, and become integrated into new units, new individuals. David Queller, in a review of Maynard Smith and Szathmáry's book, distinguished two forms of these transitions (Queller 1997). The evolution of multicellular organisms from single-celled organisms is a *fraternal* transition, because it is like an alliance of siblings. When cells produced by division of a single-celled organism do not separate but stay fused and begin to operate as a unit, their cooperation can be seen as an extreme case of kin selection. In contrast, the evolution of the eukaryotic cell by the symbiotic fusion of several prokaryotic cells was an *egalitarian* transition, because it involved entities with different capacities coming together for mutual benefit. Here the idea of reciprocity is more relevant. In homage to an earlier paper by Hamilton, Queller called his review "Cooperators since Life Began."

BOX 8.1. MODELING THE EVOLUTION OF SOCIAL BEHAVIOR

Suppose a population contains two types of individuals, *A* and *B*. Each individual interacts with a limited number of others, the individual's *neighbors*. Suppose first that each individual interacts with one neighbor only. The relationships between the fitnesses of different kinds of individuals in different neighborhoods can be written as follows.

$$(1) \quad \alpha_i = z - c + bi$$

$$\beta_i = z + bi$$

The symbols α and β represent the fitnesses of *A* and *B* individuals, respectively, and the subscript *i* describes those individuals' neighbors, with $i = 1$ for an *A* neighbor and $i = 0$ for a *B* neighbor. So α_1, the fitness of an *A* individual whose neighbor is an *A*, is $z - c + b$, and so on for the other cases. On the right-hand side, *z* (which is positive) is a "baseline" fitness, the same for everyone; *b* is the effect of having an *A* neighbor; *c* is the effect of being of the *A* type oneself. Both *c* and *b* could be positive or negative, but assume first that both are positive, so the *A* type pays a direct *cost*, and *b* is a *benefit*. Suppose there is a large population, reproducing asexually with discrete generations (Box 3.1) and where offspring are of the same type as their parents. The assignment of neighbors is done afresh in each generation. If *c* is greater than *b*, the *A* type is doomed (given that so far we are assuming interaction with one neighbor only). When

b is greater than c, the A type *may* prevail, even though it pays a cost that B does not.

Suppose that p is the overall frequency of the A type in the population at some time, and $(1 - p)$ is the frequency of B. If individuals interact at random, the chance of having an A neighbor, for both types, is just the frequency of A in the population, p. Then the *average* fitness of the A type, or W_A, is

(2) $W_A = p(z - c + b) + (1 - p)(z - c) = z - c + pb$.

Similarly, $W_B = z + pb$, and this must be higher than W_A. But it may be that the frequency of A neighbors *experienced* by the A type, or p_A, differs from the frequency of A experienced by the B type, or p_B. Then when the average fitnesses of the two types are calculated, as in (2), p_A is used rather than p. So $W_A = z - c + p_A b$, and $W_B = z + p_B b$. Now it is possible for W_A to be higher than W_B, when

(3) $p_A - p_B > c / b$.

On the left-hand side, $p_A - p_B$ measures the tendency for the A type to experience more A neighbors than the B type does. This is a measure of correlated interaction, or assortment of types. It might have a constant value, or it might vary with p. On the right-hand side, we have the relation between the cost of being an altruist and the benefits of having an altruistic neighbor. If $p_A - p_B$ is zero, we have random interaction and the A type cannot prevail. When $p_A - p_B$ is one, we have total assortment (each type interacts only with itself) and given that we're assuming b is larger than c, A must have higher fitness. When the degree of correlation is between zero and one, the fate of altruism depends on the exact relation between c and b. It is also possible for $p_A - p_B$ to be negative, in which case "spiteful" behavior can be favored, behavior in which b is negative and A pays a cost to harm others. If b is negative, the requirement becomes: $p_A - p_B < c / b$.

Formula (3) is similar to "Hamilton's rule," which was developed to describe interactions between biological relatives (Hamilton 1964, Queller 1985). Hamilton's rule has it that altruism prevails if $rb > c$, where r is the degree of genetic relatedness between individuals and b and c are, as they are here, benefits and costs associated with altruism. There are many formulas in models of social evolution that have the general form of (3), relating the costs and benefits of a social behavior to the ways that the benefits (or harms) stemming from that behavior are steered toward some individuals and away from others.

This framework can be used to represent some game theory models: as long as z, b, and c are positive, and b is larger than c, the relationships in (1) yield a Prisoner's Dilemma. If we set $z = 2$, $b = 2$, and $c = 1$, the result is the Prisoner's Dilemma discussed in the text. (There are also Prisoner's Dilemmas that cannot be represented in this way.) Imagine we have a population containing cooperators (A) and defectors (B) playing one-shot Prisoner's Dilemmas using the payoff values above. Cooperation has higher average fitness as long as $p_A - p_B > 1/2$. The model can also apply to cases where individuals have many neighbors who affect their fitness—in formula (1), the variable i may take on values other than zero and one. Then new possibilities arise. It might be that adding each successive altruistic neighbor has less and less effect—the benefit "saturates"—or the opposite might be true, so altruism needs a critical mass to be effective. To represent this we can replace (1) with

$$(4) \quad \alpha_i = z - c + bi^k,$$

$$\beta_i = z + bi^k.$$

When $0 < k < 1$, there is decreasing marginal return from adding altruists to a neighborhood; when $k > 1$, there is increasing marginal return (Godfrey-Smith and Kerr 2009). When $0 < k < 1$ and there is a reasonable degree of correlation ($p_A - p_B$), there can be situations where each type is favored when rare. When $k > 1$ each type can be favored when common, such that altruism needs a "critical mass" to be effective.

The *Stag Hunt* was introduced at the end of section 8.1. This game is usually seen as a model of a one-time mutualistic interaction, but an Iterated Prisoner's Dilemma in which the only options are TFT and ALLD can be represented as a single Stag

Hunt if played over a suitable number of trials (three or more trials for the Prisoner's Dilemma payoffs in the text). Cooperation in a one-time Stag Hunt and TFT in an Iterated Prisoner's Dilemma (with a suitable number of trials) can both be favored over defection in a situation of random interaction if they are sufficiently common, not if they are rare. In the case of the Prisoner's Dilemma in the text played over three trials, the threshold frequency of TFT beyond which it is favored over ALLD is 1/2. If the number of trials is higher, this threshold frequency is lower.

8.2. COOPERATION IN HUMAN SOCIETIES

One tradition in economics, parts of philosophy, and other fields sees human beings as fundamentally self-interested. The helping of others is an obvious empirical fact, of course, and any such view has to explain how it arose. All the models discussed in the previous section are relevant, but the details of the explanation are much affected by the question of whether altruism is psychologically shallow in humans, or runs deeper.

Our picture of this part of human psychology has been changed by a series of experiments looking at how people actually behave in game-theoretic situations in which altruistic behavior is an option. A central example is the *Ultimatum Game* (Güth et al. 1982). A sum of money is made available to two people, where one person, the "proposer," can propose a division of the money and the other, the "responder," can only accept or reject the proposal. If the division is accepted, both people get those shares; if it is rejected, no one gets any of the money.

From the viewpoint of rational choice and assuming self-interest, the proposer should offer a bare minimum and the responder should accept any share, no matter how tiny. But in experiments proposers tend to be generous, offering a third to a half of the money, and if they are less generous the responder often rejects the offer. This applies in anonymous "one-shot" situations where there is no possibility of reciprocity developing over time.

131

The experiment has been studied in many variants and many societies (Henrich et al. 2004, Bowles and Gintis 2011).

It might seem that people in these experiments are being irrational, or mistake one game for another. The idea that people do not understand the situation has been checked repeatedly, and the view that these behaviors are "irrational" is question-begging. The behaviors are irrational if all people want is their own material benefit, but the other possibility is that people have *social preferences*, preferences that involve what happens to others. They will reduce their payoff to help others, and also to punish others for violations of fairness and helpfulness. Rational choice theory does not constrain what you have preferences about; it constrains how your various preferences, along with your beliefs and actions, are related to each other. There is nothing stopping a rational individual from having preferences about the welfare of others. (It is not true that because it is *your* preferences that are being pursued, it is a matter of self-interest when you pursue them. Self-interest has to do with what those preferences are *about*.) It is an empirical fact about humans across a great many cultures that one thing people value is their own welfare, and another thing they value is helping or harming others.

Rational choice theory does not constrain what you have preferences about, I said, but evolutionary theory might seem to do this. If some people have a tendency to give resources away to others, they will lose out in an evolutionary competition to those who do not, unless some mechanism is in place that directs the benefits toward other prosocial individuals. We saw earlier in this chapter that there are various ways this can happen. The kind of prosociality that is apparent in humans points toward some explanations and away from others, though. Humans are not prosocial only to relatives, and—although the extent of this is controversial—they are prosocial in situations where it is clear that generosity will not be reciprocated. For these reasons and others, some work in this area has argued for the importance of *group*-level competition in explaining human sociality.

I'll describe one such view. The *Pleistocene* is the period of about 2.5 million years prior to the development of farming and settled human communities about 12,000 years ago. The species

Homo sapiens itself evolved in this period, roughly 200,000 years back (though problems with recognizing transitions between species over time, discussed in the previous chapter, certainly arise here). During the Pleistocene it is likely that humans lived in small communities that were very egalitarian (or rather, they were egalitarian within each sex; relations between the sexes may well have been a different matter). These communities competed with each other, both directly in warfare and less directly in their attempts to make use of resources in a difficult environment. Societies with good social cohesion, with habits of helping and norms discouraging exploitation, were effective in this competition and survived while other groups did not. Cohesion in these societies was due to a combination of traits with different origins. These include psychological features such as the social emotions of shame and pride, which evolved by natural selection, along with culturally transmitted habits and institutions. The psychology that came out of this period was one featuring strong social preferences bolstered by emotions, and a tendency to internalize local norms about proper behavior.

The view I just sketched is due especially to Bowles and Gintis (2011), and complementary views have been developed by Boyd and Richerson (2005) and others. This picture is controversial and there are plenty of rivals. One alternative view is that mutualistic cooperation was a more important factor in early human societies than views like that of Bowles and Gintis suppose (Tomasello 2009, Sterelny forthcoming). Perhaps many situations had the rough form of a Stag Hunt, where there is little temptation to exploit cooperators, and the problem to be overcome is achieving coordination. If this is right, there need be no special role at this early stage for group-level competition. Perhaps group-on-group competition, and patterns of behavior in which sacrifices are made for group benefit, came later, as societies moved toward farming and away from the simple social structures of the Pleistocene.

These debates about human groups connect, once again, to questions about levels of selection and Darwinian individuals. Clearly it makes sense to see groups as competing units in some situations, such as wars. The relation this competition has to evolutionary processes is not straightforward, however. Groups can *affect* evolution by being part of the social context of individuals, without being *units of selection* in their own right. In earlier chapters I emphasized reproduction in my account of natural selection and the levels at which it operates. In hypotheses about group competition in human evolution, what people generally have in mind is survival and growth: some groups flourish and some perish. Reproduction by human groups can happen, when a group splits or sends out colonizing parties, but reproduction is less central here than in ordinary biological evolution.[2]

In the earlier chapters I also emphasized the role of reproduction in *origin explanations*, explanations of how new traits and new kinds of life come to exist. In explaining human cooperation, questions about originations are less pressing than they are in standard biological cases, such as the evolution of the eye. Once individual humans are smart, they can come up with all sorts of new behaviors; the harder problem with prosociality is explaining how it is maintained. Explanations for the maintenance of prosocial behavior can involve fine-tuned combinations of mechanisms. In some game-theoretic situations, several outcomes can be stable equilibria, but some equilibria are better places to end up than others. Then if a range of human groups face the same kind of situation, some groups may end up at one equilibrium while others end up at another. In the Iterated Prisoner's Dilemma and in the Stag Hunt, both cooperative and noncooperative behaviors can be stable against invasion by the other within a randomly interacting group. Suppose in an Iterated Prisoner's Dilemma situation that one group reaches an ALLD equilibrium and other reaches a TFT one.[3] Each outcome is internally stable, but the more prosocial group might be more resistant to challenges from outside, and more effective in conflict between groups. Here an explanation of the maintenance of a trait is subtle and involves

[2] A comparison might be made to "modular" organisms like trees, which can in principle live and grow indefinitely, as a society can (Jackson et al. 1985).

[3] TFT is not stable against invasion by ALLC (unconditional cooperation), though ALLC can invade TFT only by random "drift," not by selection. So TFT is not an ESS in a context in which ALLC is among the options. See note 1 of this chapter. Here I also assume that the number of trials is sufficient for TFT to be advantageous.

processes at different levels, though it is easy to come up with the trait initially. Evolving an eye with a lens is a different matter; it is very hard to invent, not so hard to keep around.

Another feature of recent work in this area (whether groups are seen as units of selection or not) is the importance of *social learning*, learning that makes use of the observation of others' behavior, and in some cases demonstration and teaching. As well as enabling individuals to pick up useful tricks, this enables the transmission of ideas and skills over generations by non-genetic means (Tomasello 1999, Sterelny 2012). As discussed in chapter 4, behavioral shifts that are not due to genetic mutation can affect genetic evolution. These behavioral shifts can change the physical environment, leading to new selection pressures, and can also create new requirements for effective behavior in social settings— "changing the environment" in a sense, as each individual is part of the environment of others. Social learning is an especially powerful route by which behavioral changes can have these effects, as it enables skills to proliferate and persist over long periods. A good example is cooking. Cooking probably predates *Homo sapiens*, perhaps going back over a million years (Wrangham 2010). It was invented (perhaps a number of times) and passed on culturally by some combination of observation, copying, and teaching. Once established, cooking affects genetic evolution, because our diets are changed so much. It probably affected the evolution of our digestive system, and also our senses of taste and smell, as cooked food is safer from bacteria and other contaminants than raw food. Other distinctive features of human behavior, including language, are likely to have arisen through long-term interactions between genetic and cultural change.[4]

The later stages in these tales about human evolution include the shift around 12,000 years ago to the development of agriculture, leading to an enormous growth in population size and social complexity. The period of early human evolution in the Pleistocene looks to some like an egalitarian interlude between the strongly hierarchical social worlds of nonhuman primates

[4]See Durham (1992), Deacon (1998), Jablonka and Lamb (2005), Tomasello (2009).

and a return to hierarchy with the rise of settlement and farming. Finally we see the appearance of modern societies like our own. Standard explanations of economic arrangements in this last setting emphasize self-interest and the "invisible hand" of the market, a setting in which we make exchanges for mutual gain (Smith 1776). Bowles and Gintis, whose ideas about biology I have drawn on in this section but whose background is economics, think this standard picture is not accurate. Given the complexity of modern social life, most contracts could not really be enforced, and the invisible hand is not what keeps things going. Much instead is due to basic human prosociality, the fact that we evolved to take an interest in others and to internalize norms of fairness that play a strong motivating role.

8.3. Cultural evolution

The evolution of human behavior features an important role for copying and imitation, and other forms of social learning. When behaviors and ideas spread by copying, this raises the possibility of an evolutionary dynamic in the pool of ideas and behaviors themselves. A strong version of this view, defended by Dawkins (1976), Dennett (1995), and others, holds that cultural change is evolution in a domain of *memes*, or cultural replicators. Ideas, behaviors, and artifacts compete in an environment made up of human brains and social life. How does this view relate to the ideas above?

The requirements for evolution by natural selection are abstract. Any entities that reproduce can evolve by this process. Reproduction need not involve faithful copying or replication, as long as there is *heritability*—a tendency for parents to resemble offspring (§3.1). Faithful replication is a special case. In at least some situations it is possible for things like ideas and behaviors to be units of selection in this sense—Darwinian individuals. What is needed is that one idea or other cultural item be derived from a small number of others, its parent or parents, with similarity across the generations. It is hard to work out exactly what the relation amounting to "reproduction" here involves, but examples show it can happen in principle. Suppose in a maritime culture that each new boat is made by choosing a single existing boat and making a copy of it. Then each boat will have a unique parent boat. There might be both "differential fertility" and "differential mortality" of boats. If a boat sinks, it is not around to be copied, and it may also be that boats found to be stable or easy to sail are copied more than those that are not.[5] Small variations are introduced by accident from time to time, and change can involve the accumulation of subtle improvements.When the entities being copied are ideas rather than artifacts, the process is not so clear. People often treat ideas as definite objects when writing about this topic. There is probably a good deal of idealization going on there, but perhaps it is a reasonable way of looking at some cases, and it seems that there can at least be an approximate version of a parent-offspring relation between an idea in one person and an idea in another. If people are too smart and too flexible in how they react to the ideas of those around them, however, then the parent-offspring relationships are lost (Sperber 1996, 2000). Once people attend to many sources of information, and process what they find in intelligent ways, each new belief or artifact is *influenced* by many precursors without being the *offspring* of any of those influences. There are also forms of social learning that do not generate parent-offspring relationships even in their simple versions. An example is conformism, copying whatever is common around you (Richerson and Boyd 2005). Perhaps ideas and behaviors are acquired in human cultures by an ever-shifting mixture of copying the locally successful, copying whatever is common, obeying authority figures, and individual inventiveness. As I said in section 3.4, processes in which there is variation, recurrence, and change can be more similar to biological evolution or less so. When copying the successful has the most influence in a culture, the pattern of change shifts toward a more Darwinian one. When other habits and factors operate more strongly, the pattern is less Darwinian.

---

[5]This example is discussed in Godfrey-Smith (2012), drawing on Rogers and Ehrlich (2008).

One departure from a Darwinian pattern mentioned above involves the role of authority figures, who can impose ideas on others regardless of their consequences. This is one way in which a society can be a more *organized* system, one where the details of the relations between parts are very important, as opposed to a more *aggregative* one (§2.2). Highly organized systems can undergo adaptive change in ways that are different from evolution in populations. In a biological population, adaptation arises by the proliferation of successful variants, and this happens because successful individuals make more individuals who tend to share their traits. In individual learning, in contrast, though adaptation can arise by the retention and refinement of useful variants, this is not in general because these variants make more of themselves. Instead the system as a whole is smart enough to track success, retaining good ideas and reproducing behaviors that work (§3.4). Insofar as cultural change involves variation and selection—a controversial matter—it often seems to lie between these poles, with localized evolution by copying in some situations and more organized patterns of change in others.

Looking back at Figure 3.1 on page 29, it is interesting to ask how cultural change compares to biological evolution. In both cases one can zoom in and out, finding different relationships visible at different scales. In biology, change at larger scales occurs by the aggregation of many localized births, lives, and deaths, with reproduction at the lower level taking a regular pattern. In culture, some low-level influences are local in extent while others extend broadly, and the patterns by which one person's behaviors affect another's can change rapidly. Zooming out still further, to the kind of pattern represented in Figure 7.1, some cultural processes in human history have a tree-like pattern, with steady divergence of lineages, while others have different patterns of connection.[6]

---

[6]See Gray et al. (2007) for discussion of trees and other structures at a macro-scale in culture.

8.4. POPULATION THINKING AND HUMAN NATURE

This final section ties together the present chapter and the previous one. Chapter 7 was about species and kinds. One species of particular interest is our own, and this brings up questions of *human nature*.

The idea of human nature is a point of contact between scientific investigation of our species and many theoretical projects outside biology. How does an evolutionary perspective relate to this idea? One possibility is that an evolutionary orientation to psychology can tell us what human nature is and why it is that way (Pinker 2002). On another side, here is biologist Michael Ghiselin: "What does evolution teach us about human nature? It tells us that human nature is a superstition" (1997, p. 1).

Part of this disagreement involves stronger and weaker conceptions of what sort of thing human nature is supposed to be. On a traditional and strong conception, human nature is a combination of properties that is *universal* in humans and *distinctive* to them. These include features of our thought, behavior, and physical form. Those observable features are manifestations of a set of internal properties, also shared across all humans. (Perhaps severely handicapped humans lack some of these properties, but they are present in all "normal" humans.) Human nature, on this first view, is stable over time and hard to interfere with. Facts about human nature are relevant to moral discussion, at least by telling us what can be changed easily, and perhaps in determining what is natural to us in the sense of what is proper or appropriate. That is the sort of thing people like Ghiselin and David Hull (1986) think is a myth.

But how much of a myth? *Homo sapiens* is an easily recognized species, and once you know that someone is a human you can make predictions about him or her. These observable features are caused in large part by a genetic profile that is common across humans. If you want to know why humans look so unlike chimps and sturgeons, DNA is not the whole story, but it is the most important difference maker (§6.2). If Martians came down and needed a *field guide* to the animals found on earth, there could be a useful field guide entry for our species—bipedal, relatively

139

hairless, sociable, talkative.[7] The Martians could recognize us by how we look and what we do. In that sense, there is surely nothing mythical about the idea of human nature.

In sorting these issues out I will use a distinction, from the work of Ernst Mayr, between *typological thinking* and *population thinking* (1959).[8] Typological thinking sees variation within a species as imperfection in the worldly realization of an ideal "type." Mayr traced these views back to Plato. Population thinking inverts this perspective: nature contains populations of unique individuals, and types are rough conceptual tools we use to get a handle on this complexity. Mayr claimed that Darwin "replaced" the typological attitude in biology with population thinking. Historians have criticized this as a caricature of pre-Darwinian thought. Still, Mayr's notion of "population thinking" summarizes an outlook that was at least much strengthened by evolutionary theory, fits well with it, and motivates a shift in thinking about many topics.

The nature of human beings is one of these topics. As discussed in the previous chapter, the whole idea of species is problematic in some ways, especially when comparing organisms alive at different times, but if we set that problem aside for the moment there is nothing problematic in talking of the "nature" of the human species in a low-key way. As a result of our evolutionary history, there is a genetic profile that is characteristic of our species, which includes important causes of many of our distinctive traits. There is a temptation to see the features common to humans as a "nature" in a more fixed sense, but evolution is open-ended. The profile that applies to humans now is probably changing. New variations appear. Most are weeded out. The ones not weeded out are "abnormal" initially, but the abnormal today can become normal tomorrow. The traits that make humans distinctive now began as rare abnormalities in populations that looked very different. The typological mind-set has it that variation *within* a type reflects imperfection or abnormality, and attention is focused on

---

[7] For this analogy see Machery (2008), who acknowledges Paul Griffiths.

[8] Mayr's distinction is philosophically controversial as well; see Sober (1980), Lewens (2009), Ariew (2008), Hey (2011). For the history, see Winsor (2006), McOuat (2009).

the characteristic differences *between* types. But evolution by natural selection is, in effect, a machine that turns the former kind of variation into the latter. The differences between "types" of organisms have their origins in variation within populations, filtered and magnified in a way that yields large-scale change.

A picture that has often been attractive is that in a species like ours, there is a set of stable features set by internal causes, and the environment perturbs these features and introduces variability. By means of learning and related kinds of sensitivity, the idiosyncrasies of an individual's circumstances leave their mark. A debate then arises about which of our features are due to the stable internal causes and which are due to the environment (nature versus nurture). Learning does have an evolutionary function of fine-tuning the behavior of organisms to the details of their circumstances. But the sociality of the human species interferes with this association between internal causes and stable features, on one side, and external causes and variability, on the other. Human life is characterized by reliably transmitted behaviors and practices, making use of teaching and behavioral modeling by parents and others. As a result, many of our traits are learned, but recurring rather than idiosyncratic, and this may be integral to the evolution of many of our distinctive features as a species—as expressed in the title of Kim Sterelny's book *The Evolved Apprentice* (2012). Furthermore, traits arising from internal causes are often associated with a resistance to change, but *reliably appearing and strongly influenced by genetic causes* does not imply *hard to change*. An environmental influence that would change a trait might be rare in natural circumstances, but easy to bring about once we know what that influence is.

Views of human nature provide a framework for many discussions of moral issues. It is possible to treat human nature as bad, as in the Christian tradition, but it is also common to look for positive guidance—for information about what is natural for us, or "in accordance with nature," in a morally relevant sense. This connects to the discussion of teleology in chapter 4. If a trait has been successful under natural selection, it will be possible to describe it in terms of its *function*, in terms of what it is "supposed" to do. But as discussed earlier, the senses of "supposed"

141

and "normal" that come out of natural selection in this way are not morally loaded senses. If some behavior has an evolved function, all that means is that it has been associated with reproductive success and has been kept around for that reason. The fact that some habit or characteristic is "natural" in this sense does not, and should not, prevent us from criticizing it and perhaps trying to change it.

So the concept of human nature that can be maintained within an evolutionary and "population thinking" mind-set has substantial differences from more traditional views. Once evolution in a lineage has actually taken a particular path for a while, we can talk about an "evolved nature" that has been established in that lineage, though much of it will not be universal, even at a time. As evolution is open-ended, this talk about our nature has a *post hoc* character. A new characteristic that is "abnormal" now might be the basis for a new nature in the future. That much is true of all species, not just humans. The capacities for learning and cultural transmission seen in humans give this evolutionary openness an extra dimension.

Looking back at the history of philosophy, has anyone had a view of the status of human nature like this? Ideas that look like this were arrived at, on a very different road, by some of the 20th-century existentialists. Jean-Paul Sartre claimed that there is no human nature that does or should constrain the actual facts of human behavior and choice. Humans are what they make of themselves. Just as I put it above, talk of human nature is post hoc. To use the existentialist terminology, *existence* (the actual events of human life) precedes *essence* (any inscribed nature or "conception" of man). Existentialists produced rather metaphysically convoluted ways of expressing this point, but I think they did glimpse something important.

> When we think of God as the creator, we are thinking of him, most of the time, as a supernal artisan. . . . Thus, the conception of man in the mind of God is comparable to that of the paper-knife in the mind of the artisan: God makes man according to a procedure and a conception, exactly as the artisan manufactures a paper-knife, following

a definition and a formula. . . . In the philosophic atheism of the eighteenth century, the notion of God is suppressed, but not, for all that, the idea that essence is prior to existence; something of that idea we still find everywhere, in Diderot, in Voltaire and even in Kant. Man possesses a human nature; that "human nature," which is the conception of human being, is found in every man; which means that each man is a particular example of a universal conception, the conception of Man.

What do we mean by saying that existence precedes essence? We mean that man first of all exists, encounters himself, surges up in the world—and defines himself afterwards. (Sartre 1946/1956, p. 349)

FURTHER READING

For cooperation and altruism, Hamilton (1998), Sober and Wilson (1998), Kerr et al. (2004), Skyrms (2004), Nowak (2006b), West et al. (2007), Calcott (2008), Harman (2011); on human sociality, Tomasello (2009), Seabright (2010), Kitcher (2011); on social learning, Heyes (2012); on cultural evolution, Hull (1988), Mesoudi et al. (2006); on human nature, Oyama (2000), Dupré (2001), Pigliucci and Kaplan (2003), Buller (2005), Machery (2008), Downes and Machery (2013), Prinz (2012).

Information

In his book *Natural Selection* (1992), George Williams claimed that there are two "domains" in which biological change occurs. One is material and the other is "codical," a domain of information. In evolutionary processes information is created, persists, proliferates, and is lost.

Initially it seems that information exists only where there is something like communication or thinking going on, and although some parts of biology cover these phenomena, most do not. However, over the past half century biology has become drenched in informational terminology and theoretical ideas. Genetics is about coding, translation, and editing. In developmental biology, chemical gradients provide "positional information" to the developing organism. Biology, for many, has become a science in which information occupies a central place.

I'll argue in this final chapter against some of the most strongly information-infused views of biology. Then, however, I'll look at the unifying role of a related idea: communication.

9.1. Information and Evolution

This book has taken a straightforwardly materialist view of living systems and their evolution. Organisms are complex material objects, and the metabolic processes characteristic of life are physical processes. Some organisms live longer and reproduce more than others, where reproduction is the making of a new material object. In biological systems material things come and go from the world, use energy, change, and give rise to new material things. Evolution by natural selection is one aspect of this great array of physical goings-on.

For Williams and others, this underestimates the role of information and fails to recognize the special status of genes. "A gene is not a DNA molecule; it is the transcribable information coded by the molecule," "the gene is a packet of information, not an object" (Williams 1992, p. 11). Richard Dawkins describes evolution as flow in a river of information, a river that "flows through time, not space" (1995, p. 4). Why say anything like this? I accepted in chapter 6 that cells contain code-like structures and undergo processes that are similar to computation. But those structures and processes are found within cells, located in space as well as time.

In defending the idea that a gene is a piece of information rather than a material object, Williams suggested that genes are analogous to books, such as *Don Quixote*. A book persists as an object across many copies in different media. A DNA sequence, similarly, persists over many copyings, changes in its matter. A gene and a book are not *that* similar, as it is an important fact about DNA that its information does not so readily move from medium to medium. Some DNA sequences give rise to partially corresponding sequences in proteins, but that is a one-way street. The special features of *Don Quixote*—which is lost from the world only if all the books *and* all the e-copies *and* the microfilms, and so on, are gone—do not apply. Furthermore, the importance of thinking of *types*, as well as instances or tokens, does not only apply to DNA. Proteins, sugars, and lipids arise in many instances or tokens, but this does not make them anything other than material objects. The way that molecules of DNA usually come into existence is unusual, by copying preexisting molecules in a way that forms lineages. But this is another material fact, a fact about what happens in some places and not others and with some material substances and not others.

Another motivation for the view that evolution is an informational process comes from the idea that an evolving population accumulates information about its environment. For Dawkins, a species is a computer that "builds up, over the generations, a statistical description of the worlds in which the ancestors of today's species members lived and reproduced" (1998, p. 239). It is true that evolution is a process in which earlier events leave marks and traces that are present later. That itself is nothing unusual in

145

a physical process. A geological formation, such as a mountain, contains traces of the processes that produced it, in its rock strata and other features (including its fossils). Changes in gene pools have causes, and sometimes it is possible to work out, within limits, how a species reached its present state. The sequence of branching events in the tree of life leaves marks from which the history can be reconstructed. When the past leaves traces in the present in this way, these traces in a sense are "signs," but only in the way seen also in ordinary tree rings, which can be used to infer the history of a tree but have no further role. So far there is no reason to think that evolution has a relationship to information that other physical processes do not have.[1]

Is that all there is to say? Even if the strongest claims about the link between information and evolution are rejected, there seems to be something important here. The organisms resulting from the evolutionary process seem to benefit, in terms of adaptation, from the effects of past environments on their gene pool; geological processes don't include anything like that. And I said myself in chapter 6 that DNA is a kind of memory. If so, what is being remembered? To sort these things out I will go back to basics.

9.2. Senders and receivers

This section discusses two models—or more exactly, a model plus a framework that began within another model, but has become broader in its application. I'll begin with the latter, which is *information theory*, or the mathematical theory of communication, developed primarily by Claude Shannon (1948).

Shannon set out by imagining the transmission of a message over a channel. At one end there is a *source*, some aspect of the world that can be in a number of different states. A *transmitter* generates a *signal* that can be sent over the channel, and some

---

[1] This role for information as a theoretical concept in biology should not be confused with the growth of "bioinformatics," which is concerned with the effective use of information technology (computers, data storage, search methods) in the study of biological systems.

agent at the other end uses the signal to reduce their uncertainty about what is happening at the source.

Shannon's framework was introduced by imagining agents sending and receiving messages, but it applies also to cases where no agents play these roles in any recognizable form. The state of the weather in New York City is a *source* because it varies from day to day. The state of the clouds over New York carries information about the weather to come, because it reduces uncertainty, to some extent, about that weather. Whenever there are two variables that take on different values, and the value of one is associated with the value of another, the first variable carries information about the second. The second also carries information about the first—this relationship is symmetrical.

The *amount* of information associated with a variable is a measure of how much uncertainty it embodies, and the *mutual information* between two variables is a measure of how well you can predict the state of one from the other. The formulas for these are given in a note.[2] Philosophers have sometimes called predictive relationships of this kind *natural meaning* or *indication* (Dretske 1988). Information in this sense is all over the place and has nothing special to do with evolution. If you can make inferences about past pressures of natural selection from present gene frequencies, then the gene pool contains information about the past. If you can infer the geological history of a cliff from its rock layers, those rock layers also carry information about the past. This is just a way of describing dependence relations between states of the world that arise from physical connections, direct or indirect, between them.

I said that Shannon's model was set up by imagining something like a sender and receiver, but those roles are not essential to the resulting framework. Now, though, let's look more closely at those roles. The second model I will describe was developed

---

[2]The *Shannon entropy* of a discrete random variable X which has possible values $x_1, x_2, x_i \ldots$ is $H(X)=-\Sigma P(x_i) Log_2 P(x_i)$, where $P(x_i)$ is the probability of the value x_i. The *mutual information* between two variables X and Y is $H(X) + H(Y) - H(X,Y)$. Here $H(X,Y)$ is the "joint entropy" of X and Y, the entropy of the distribution of combinations of X and Y values. The mutual information between two variables is zero when they are completely independent.

by David Lewis (1969), who wanted to understand human communication. Lewis imagined a *communicator* and an *audience*, though I'll shift the terminology to *sender* and *receiver*. Lewis took for granted that the sender could get messages of some kind to the receiver if they choose to, and wanted to understand how and why these behaviors come to exist. In effect, Lewis was analyzing what Shannon took for granted, and vice versa.

In Lewis's model he assumes that the sender can see the state of the world, but cannot act except to produce signals. The receiver can only see these signals, but can act in a way that affects them both. Lewis assumed *common interest* between sender and receiver: the two agents have the same preferences for what they want done in each state of the world. It is also assumed that a receiver's act that works well in one state of the world does not work in others. In this situation, a division of labor is possible; the sender acts as eyes, the receiver as hands. If the sender comes to send distinctive signals in each state of the world, and the receiver uses these signals to prompt the appropriate action in each state, then this sending and using of signals is a *Nash equilibrium*— neither side has any incentive to change (unilaterally) what they are doing.

Lewis assumed that the sender and receiver are intelligent agents who make choices about these behaviors. Brian Skyrms (1996, 2010) recast the model in an evolutionary framework, showing that sending and receiving behaviors of this kind can also evolve by natural selection and be evolutionarily stable.

Mutual information arises all over the place, as we saw, as a result of ubiquitous physical and chemical processes. But the Lewis model describes one way that it can come to exist, by the shaping of senders' behaviors. If a sender can see the world and can produce signs, he or she has the option of creating signals that carry information about the state of the world. But why should the sender do this? In a Lewis-like situation, a sender comes to do this (through choice or evolution) because they benefit from doing so, given the receiver's rules of action. If the sender and receiver want entirely different acts performed in each state of the world, the receiver could use informative signals to exploit the sender, and the sender would have reason to stop sending them; in most cases (though not always) stable communication cannot then be sustained. The Lewis model, especially in the version developed by Skyrms, describes how the production of informative signals and their use *as* signals in the guidance of behavior coevolve

Many natural processes have some fit to this model, though sometimes a very partial fit. The model can be applied to signaling between organisms and within them, and also to cases where the boundaries of the organisms are not clear. Signaling can occur across space and across time. In honeybees, a worker who finds a source of nectar will perform a "dance" on returning to the hive, which carries information about the distance and direction of the nectar source (Von Frisch 1993). Here the sender has information the receivers do not have, as in the Lewis model, but the sender can act (fetching nectar) as well as signal, and the point of signaling is to recruit *more* carriers. The fit to the model is not complete, but fairly good. Animal alarm calls are a more controversial case. In many group-living animals an individual who spots a predator will give a call, sometimes also giving information about the kind of predator. In at least some cases this is a cooperative, perhaps an altruistic, behavior, though it is often hard to tell the real costs of alarms and whether the benefits of calling tend to fall especially on other callers, such as the caller's kin. Another intriguing case is signals that appear to be given by some prey animals *to* predators, which may be "I've seen you; don't bother" signals. Predator and prey have opposing interests to a large extent, but there is some overlap; both will prefer to avoid a chase in a situation where the prey animal is sure to get away.[3]

Lewis's original target was human language. Real speakers and hearers are much more complicated in their behaviors and agendas than the model allows, but it may be that linguistic communication has a cooperative core that the model does capture (Millikan 1984, Harms 2004, Tomasello 2008). Applications to interactions within organisms raise different complications. An obvious-looking example is signaling between neurons in the

---

[3] For alarm calls see Cheney and Seyfarth (1990); for both cases of animal communication discussed here see Searcy and Nowicki (2006) and Bradbury and Vehrencamp (2011).

brain, but except in very simple nervous systems, a single neuron can't produce actions of a kind that have consequences of the kind the model assumes (Cao 2012). Behavior is produced by many neurons working in concert. The idea of "common interest" might also seem questionable when applied to parts of a single organism, but all talk of "interest" here should be seen as a shorthand. What is essential is the presence of some process of selection by which the consequences of sender and receiver behaviors stabilize or reshape those behaviors. In this book we've seen several times that there is a family of processes that can play this role, including evolution by natural selection, learning by reinforcement, some cultural processes involving imitation, and deliberate choice (§§3.4, 4.3, 8.3). The terms "sending" and "receiving" are to be understood very broadly here too, to include various kinds of writing and reading, producing and consuming, marking and interpreting.

Another kind of partial case, or a fragment of what is covered by the Lewis model, comprises situations where an informative sign has no sender, but is used by a receiver or interpreter. In biology, the term "cue" is often used in contrast to "signal" for unsent or naturally occurring signs of this kind. Cues, like clouds as indicators of rain, can be used in much the way that signals can. Some might prefer to say that there *is* a sender in these cases, but a special kind of sender who will keep sending informative signals no matter what the receiver does. However they are described, there is a difference between cases where sender and receiver behaviors can both change as a consequence of the pairing of the receiver's actions with states of the world, and cases where one side or the other is unaffected by the consequences of those actions.

Putting some of those comparisons together, we can say that sender-receiver systems appear in *clear* cases and more *marginal* ones. In the clear cases, distinct objects play the roles of sender, sign, and receiver, and a selection process of some kind stabilizes the sender's and receiver's behaviors. The signs that mediate between them carry information about the world, and do this *because* their sender's behavior has been shaped by a selection process to produce such signs. The signs have an effect because the receiver's behavior, too, has been shaped by selection. In more *marginal* cases, a system has only a loose fit to those roles—the sign and the reader or receiver might not be distinct, for example, the signs might not be produced by a sender whose behavior has been shaped by selection, or the interactions covered by the Lewis model are largely submerged by others.

Sender-receiver setups are so ubiquitous in everyday human experience that strong habits of description have grown up around them. We are all used to *talking about* signs, as well as playing the sender and receiver roles ourselves. This has consequences when scientists—who are people, too—encounter systems that are partial or marginal cases of the sender-receiver setup during their research into genetic systems, ecological networks, and the like. Habits of description and interpretation that have their home in our dealings with the clear cases are activated, and applied to the partial cases.

Arnon Levy (2010) has argued for a "fictionalist" view of much informational terminology in modern biology. He suggests that talk of messages and signals in areas like genetics and developmental biology does not involve commitment to an extra "domain," or show that we need a new theory of meaning that applies to tiny subpersonal systems, but these descriptions are not mere loose talk either. They involve a socially established *pretence*, in which parts of living organisms are described *as if* they are communicating with each other even though they are not. Once people are used to working within the pretence, empirically grounded discussion can go on within it—if someone says that information about which way is up is made available to the cell at stage X but not stage Y, people know what that means and the claim can be tested. But, for Levy, the message-using cell is still a fiction. The view I am suggesting here is not quite the same as this, but related, and I would draw on Levy's analysis in some cases. Lots of empirical systems have a *partial match* to a communicative setup that we are familiar with from everyday life, and that people draw on when describing what cells and other parts of organisms do. Some of these descriptions use terminology that is literally applicable only when talking about paradigm cases of human communication, but the boundary between the literal and the metaphorical in this area is unclear.

Let's now return to connections that have been drawn between evolution and information. Dawkins claimed that a gene pool is a statistical summary of the experience of a species in the past, and an organism is a "description" of the world of its ancestors (1998). This can first be understood as a matter of mutual information, in the sense of information theory. The gene pool contains marks of its past; from the structure of an organism you can predict some facts about where and how it lives, or rather, where its ancestors lived. Things like this are true, however, of rock formations as well—their past can be inferred, to some extent, from their present. The next question to ask has now become clearer: does the information about the past in an organism or gene pool have a sender and receiver (writer/reader, producer/consumer)?

An organism itself is not a message in this sense. The case of a gene pool is less straightforward. If a gene *pool* has a reader or user, it is the whole population of organisms whose genomes are drawn from it. But this population is a collection of separate organisms in evolutionary competition with each other. It is not anything like an agent in its own right. Maybe the user of the message is an individual organism? All it can read is its own genome, though, not the gene pool as a whole. I said in chapter 6 that a genome is like a library, and I'll return to this in the next section. For now, my earlier point seems intact—evolution itself is not an information-using or information-involving process in a way that marks it off from other processes of change. There is a lot of information in Shannon's sense around, just as there is with other processes in which what happens at one place and time leaves marks at other places and times.

I have been arguing that evolution is not different from other physical processes with respect to its relationship to information. There are some who think that physics is teaching us that information is at the bottom of *everything*. Not only the living world, but the physical universe has an informational nature. Perhaps this is true, but I think some of the ambiguities that arose above—between the mere presence of information in Shannon's sense and stronger hypotheses of information *use*—are relevant here too.

9.3. COMMUNICATION SINCE LIFE BEGAN?

In section 6.2 I discussed the role of information in genetics. Some recent work has explicitly approached genetic phenomena using models of the kind discussed in this chapter (Shea 2007, 2012, Bergstrom and Rosvall 2010). Looking within a cell, the "reading" of genes is a fairly well-defined matter—transcription and translation. But who makes the message that these readers read? One option (seen in Bergstrom and Rosvall) is to say this is the organism's parents. Then the offspring is the receiver, and inheritance is communication between generations. A possible problem here is a kind of equivocation over levels: if the readers are cell-level structures, then it seems the sender can't be a whole organism, a collection of many cells. In Shea's version of the view, the user of the genetic message is the entire "developmental system" in the offspring, a large collection of processes that together enable the genes to have their effects. Perhaps this is fine, and the possible problem above with Bergstrom and Rosvall's view does not look especially serious either, but it seems that this direct application of the model might be leading to some artificial forcing of entities into the model's roles.

Here is another way of looking at it. As argued in section 6.2, the "reading" of the genome occurs only at the cell level, by the machinery of protein synthesis. In that chapter I also used an old quote by David Nanney (1958), who said a genome is a "library of specificities." DNA forms a cell-level system of *memory*, one that has the further role of participating in the accessing of that memory. When the idea of DNA as memory is cast within the sender-receiver model, some interesting relationships between memory systems emerge.

Memory, from the viewpoint of the sender-receiver model, is the sending of messages across time. "Sending" is not the ideal term in this context, as something "sent" is usually out of circulation during the time it takes to get to its receiver. In memory, a representation is often available over an extended period. But that is a superficial difference. If the reader of the genetic message is the machinery of protein synthesis, though, who is the sender or producer of the message? What nature has given rise to here is something different from a clear send-receive or write-read system. The evolutionary process—which is not a sender in the sense of the model—shapes the genetic sequences that are around at any time. This occurs by mutation and differential reproduction. It was tempting there to write "the evolutionary process does this by mutation and differential reproduction," but that wording suggests an agent using a method to get something done, and evolution is not like that. Evolution just happens, and one result is that cells contain some DNA sequences rather than others. What cells do with those DNA sequences is read them, consult these libraries, in the manufacture of proteins and other gene products. The message has no sender or writer or producer, but it can be read anyway; this is an *evolve-read* memory system rather than a *write-read* memory system.[4]

I said the genetic message has no writer or producer, but part of the story here is DNA replication. Another part, in sexual organisms, is the mixing of genetic material in recombination. The first of these, replication, I treat as more akin to persistence than to writing or sending. Recombination is different, as it produces novelty. At this point I think we have reached features of genetic systems that have no analogue in the familiar send-receive systems that guided the construction of the model. Nature has given rise to something different. There is some match between basic features of genetic systems and a version of a sender-receiver system in which time is being bridged and an evolutionary process has replaced the sender, but the match is only partial.

Genetic systems can be contrasted with memory in the brain. Since the time of Plato, memory has been conceived by philosophers most often in terms of "inscription"—a write-read model of memory. From the 18th century onward, philosophers and then psychologists trying for a mechanistic understanding have

---

[4]Once we are looking at the level of cells, the idea of sending information *across* generations to a *new* organism becomes problematic. At the cellular level there is continuity of living cells, across mother and offspring. DNA persists across cell division and replication, entering new contexts. "When does human life begin? Never, for it is part of an unbroken series of generations that goes back to Darwin's warm little pond" (Ghiselin 1997, p. 1).

often avoided this model. The *associationist* tradition holds that experience leaves its mark on the mind, and this has effects later, but these "marks" should not be seen as things that are written and then read. At present there is a divergence of opinion in neurobiology and cognitive science on this issue. Computers do have a write-read memory, as illustrated in a simple but powerful form in a Turing Machine, the abstract design for a computer developed by Alan Turing in 1936. A Turing Machine has a "head" that writes and reads marks on a "tape." Many psychologists and some neurobiologists think the brain must have a memory system that is similar to this, even though it is not immediately visible (Gallistel and King 2010). More neurobiologists and some psychologists think the message of recent brain science is instead that brains have devised a different way of solving the problem of storing memories, and the write-read model does not apply (Koch 1999). This second view might be expressed by saying that rather than a *write-read* mechanism the brain uses a *write-activate* memory. The marks left by experience in memory can do their job without a reader. If this is right, the brain is a kind of flipside to the case of genetic systems, as it is the reader that has been avoided; *write-activate* rather than *evolve-read*.

With this comparison laid out it is interesting to add another memory system within cells, a system more readily seen *as* memory: the modification of DNA with chemical marks (especially methyl groups) that inhibit transcription. Some people call this an "epigenetic code." In that case the "writing" step is clear; the DNA is marked in a systematic way by machinery with that function. It might then seem that this memory system fits a write-read model, but to say this is to understand the "read" step in a broad way. Once a mark is made, the usual consequence is that this piece of DNA is *not* read. So this could also be called a *write-inactivate* memory system. Marking works as a difference maker by the prevention of reading. (Epigenetic processes also include the binding to DNA of chemicals that encourage transcription.) So within the varieties of memory there are systems that have a write-read character (which I see as fitting the sender-receiver model), and there are variants that get by without a write step or a read step, perhaps both.

I'll discuss one other application of these models. A basic feature of life in multicellular animals is *differentiation*. Nearly all the cells within you have extremely similar genomes (§5.2), but those cells differentiate into very different types (skin, brain, liver). This works by the regulation of gene expression; only a small number of genes are expressed in any given cell, and this pattern of expression is determined largely by the binding of molecules to the DNA. This involves memory, as seen above, and it involves signaling over space as well.

In development, a cell's environment is the source of molecular cues and signals. Some derive externally to the organism, from the mother or elsewhere. But as the process goes on, the cells in the growing organism also affect each other, engaging in cell-to-cell signaling. Cells emit chemicals that alter the environment of other cells, affecting their gene expression. As Ben Kerr observed, the signaling processes that play this role can be seen as a combination of *niche construction* and *phenotypic plasticity*. These were discussed earlier in this book (§§4.2, 6.2) as features of whole organisms; now they are seen between cells within an organism, and they are part of how the organism comes to be. In niche construction, a living thing alters its environment. In plasticity, a living thing generates different phenotypes as a function of its environment. In a way, all signaling is a combination of these two things: the sender transforming its surroundings, the receiver engaging in a plastic response.

There are continuities between the simple kinds of signaling seen between cells, through animal communication, to the highly elaborated forms seen in human communication—gesture, speaking, picturing, writing—that arise from the special forms of social involvement characteristic of our species. Like cooperation, communication has a dual role, both as a part of human social life and as an element in how many other biological entities come together and function. Communication-like behaviors are ubiquitous, and communication is also a manifestation of something more basic. A combination of receptivity and activity, with those behaviors stabilized by selection, by feedback, is a distinctive feature of the living world.

Further Reading

On information, see Dretske (1981), Dennett (1987), Bergstrom and Rosvall (2010), Donaldson-Matasci et al. (2010), Gleick (2011); on the sender-receiver model, Skyrms (2010); on signaling in genetic systems, Ptashne and Gann (2001); on animal communication, Stegmann (2013) and the references in note 3; also Millikan (1984, 2004), Griesemer (2000), Bonner (2000).

References

Abrams, M. 2009. The Unity of Fitness. *Philosophy of Science* 76: 750–61.

Amundson, R. 1989. The Trials and Tribulations of Selectionist Explanations. In K. Hahlweg and C. A. Hooker, eds., *Issues in Evolutionary Epistemology*. Albany: State University of New York Press.

———. 2005. *The Changing Role of the Embryo in Evolutionary Thought: Roots of Evo-Devo*. Cambridge: Cambridge University Press.

Ariew, A. 2008. Population Thinking. In M. J. Ruse, ed., *Oxford Handbook of Philosophy of Biology*. New York: Oxford University Press.

Ariew, A. and Lewontin, R. C. 2004. The Confusions of Fitness. *British Journal for the Philosophy of Science* 55: 347–63.

Armstrong, D. 1985. *What Is a Law of Nature?* Cambridge: Cambridge University Press.

Axelrod, R. 1984. *The Evolution of Cooperation*. New York: Basic Books.

Axelrod, R. and Hamilton, W. 1981. The Evolution of Cooperation. *Science* 211: 1390–96.

Ayre, D. J. and Grosberg, R. K. 2005. Behind Anemone Lines: Factors Affecting Division of Labour in the Social Cnidarian *Anthopleura elegantissima*. *Animal Behaviour* 70: 97–110.

Bateson, W. 1900. Problems of Heredity as a Subject for Horticultural Investigation. *Journal of the Royal Horticultural Society* 25: 54–61.

Beadle, G. W. and Tatum, E. L. 1941. Genetic Control of Biochemical Reactions in Neurospora. *Proceedings of the National Academy of Sciences USA* 27: 499–506.

Beatty, J. 1995. The Evolutionary Contingency Thesis. In G. Wolters and J. Lennox, eds., *Concepts, Theories, and Rationality in the Biological Sciences: The Second Pittsburgh-Konstanz Colloquium in the Philosophy of Science*. Pittsburgh: University of Pittsburgh Press.

———. 2006. Replaying Life's Tape. *Journal of Philosophy* 7: 336–62.

Bechtel, W. and Richardson, R. C. 1993. *Discovering Complexity: Decomposition and Localization as Strategies in Scientific Research*. Princeton: Princeton University Press.

Bedau, M. A. 1997. Weak Emergence. In J. Tomberlin, ed., *Philosophical Perspectives: Mind, Causation, and World*. Oxford: Blackwell.

———. 2007. What Is Life? In S. Sarkar and A. Plutynski, eds., *A Companion to the Philosophy of Biology*. New York: Blackwell.

Bedau, M. A. and Humphreys, P., eds. 2008. *Emergence: Contemporary Readings in Philosophy and Science*. Cambridge, MA: MIT Press.

References

Beebee, H., Hitchcock, C., and Menzies, P., eds. 2010. *Oxford Handbook of Causation*. Oxford: Oxford University Press.

Bennett, J. 2003. *A Philosophical Guide to Conditionals*. Oxford: Clarendon.

Benzer, S. 1957. The Elementary Units of Heredity. In W. D. McElroy and B. Glass, eds., *A Symposium on the Chemical Basis of Heredity*. Baltimore: Johns Hopkins University Press.

Bergstrom, C. T. and Rosvall, M. 2010. The Transmission Sense of Information. *Biology and Philosophy* 26: 159–76.

Binmore, K. 1998. Review of *The Complexity of Cooperation* by Robert Axelrod. *Journal of Artificial Societies and Social Simulation* 1, http://jasss.soc.surrey.ac.uk/1/1/review1.html.

Bishop, C. D., Erezyilmaz, D. F., et al. 2006. What Is Metamorphosis? *Integrative and Comparative Biology* 46: 655–61.

Blute, M. 2007. The Evolution of Replication. *Biological Theory* 2: 10–22.

Bonner, J. T. 1974. *On Development: The Biology of Form*. Cambridge, MA: Harvard University Press.

———. 2000. *First Signals: The Evolution of Multicellular Development*. Princeton: Princeton University Press.

Bouchard, F. 2008. Causal Processes, Fitness and the Differential Persistence of Lineages. *Philosophy of Science* 75: 560–70.

Bouchard, F. and Huneman, P., eds. 2013. *From Groups to Individuals: Evolution and Emerging Individuality*. Cambridge, MA: MIT Press.

Bowler, P. J. 2009. *Evolution: The History of an Idea*. 25th Anniversary Edition. Berkeley: University of California Press.

Bowles, S. and Gintis, H. 2011. *A Cooperative Species: Human Reciprocity and Its Evolution*. Princeton: Princeton University Press.

Boyd, R. N. 1999. Homeostasis, Species, and Higher Taxa. In R. A. Wilson, ed., *Species: New Interdisciplinary Essays*. Cambridge MA: MIT Press.

Boyd, R. T. and Richerson, P. J. 2005. *The Origin and Evolution of Cultures*. Oxford: Oxford University Press.

Bradbury, J. W. and Vehrencamp, S. L. 2011. *Principles of Animal Communication*. 2nd ed. Sunderland: Sinauer Associates.

Brandon, R. N. 1978. Adaptation and Evolutionary Theory. *Studies in History and Philosophy of Science* 9: 181–206.

———. 1990. *Adaptation and Environment*. Princeton: Princeton University Press.

Browne, J. 1996. *Charles Darwin: Voyaging*. Vol. 1. Princeton: Princeton University Press.

———. 2003. *Charles Darwin: The Power of Place*. Vol. 2. Princeton: Princeton University Press.

Buller, D. J. 1999. *Function, Selection, and Design*. Albany: State University of New York Press.

———. 2005. *Adapting Minds*. Cambridge, MA: MIT Press.

Burian, R. M. 2004. Molecular Epigenesis, Molecular Pleiotropy, and Molecular Gene Definitions. *History and Philosophy of the Life Sciences* 26: 59–80.

Burkhardt, R. W. 1977. *The Spirit of System: Lamarck and Evolutionary Biology*. Cambridge, MA: Harvard University Press.

Burnet, M. 1958. *The Clonal Selection Theory of Acquired Immunity*. Cambridge: Cambridge University Press.

Burt, A. and Trivers, R. 2006. *Genes in Conflict: The Biology of Selfish Genetic Elements*. Cambridge, MA: Harvard University Press.

Buss, L. W. 1987. *The Evolution of Individuality*. Princeton: Princeton University Press.

Calcott, B. 2008. The *Other* Cooperation Problem: Generating Benefit. *Biology and Philosophy* 23: 179–203.

Calcott, B. and Sterelny, K., eds. 2011. *The Major Transitions in Evolution Revisited*. Cambridge, MA: MIT Press.

Campbell, D. T. 1960. Blind Variation and Selective Retention in Creative Thought as in Other Knowledge Processes. *Psychological Review* 67: 380–400.

Cao, R. 2012. A Teleosemantic Approach to Information in the Brain. *Biology and Philosophy* 27: 49–71.

Carnap, R. 1966. *Philosophical Foundations of Physics*. New York: Basic Books.

Carroll, J. W., ed. 2004. *Readings on Laws of Nature*. Pittsburgh: University of Pittsburgh Press.

Cartwright, N. 1983. *How the Laws of Physics Lie*. Oxford: Clarendon.

Changeux, J. P. 1985. *Neuronal Man: The Biology of Mind*. New York: Pantheon.

Cheney, D. and Seyfarth, R. 1990. *How Monkeys See the World: Inside the Mind of Another Species*. Chicago: University of Chicago Press.

Clark, E. 2011. Plant Individuality and Multilevel Selection Theory. In B. Calcott and K. Sterelny, eds., *The Major Transitions Revisited*. Cambridge, MA: MIT Press.

Cook, R. E. 1980. Reproduction by Duplication. *Natural History* 89: 88–93.

Coyne, J. A. and Orr, H. A. 2004. *Speciation*. Sunderland: Sinauer Associates.

Cracraft, J. 1983. Species Concepts and Speciation Analysis. *Current Ornithology* 1: 159–87.

Craver, C. 2001. Role Functions, Mechanisms and Hierarchy. *Philosophy of Science* 68: 31–55.

References

———. 2009. *Explaining the Brain*. Oxford: Oxford University Press.

Crick, F. 1958. On Protein Synthesis. *Symposia of the Society for Experimental Biology* 12: 138–63.

———. 1970. Central Dogma of Molecular Biology. *Nature* 227: 561–63.

Cummins, R. 1975. Functional Analysis. *Journal of Philosophy* 72: 741–65.

Darwin, C. 1839. *Journal and Remarks. 1832–1836. (Voyage of the Beagle)*. London: Henry Colburn.

———. 1859. *On the Origin of Species by Means of Natural Selection, or the Preservation of Favoured Races in the Struggle for Life*. London: John Murray.

———. 1871. *The Descent of Man, and Selection in Relation to Sex*. London: John Murray.

Darwin, E. 1794. *Zoonomia; or the Laws of Organic Life, Part I*. London: J. Johnson.

Dawkins, R. 1976. *The Selfish Gene*. New York: Oxford University Press.

———. 1982. *The Extended Phenotype: The Long Reach of the Gene*. New York: Oxford University Press.

———. 1986. *The Blind Watchmaker: Why the Evidence of Evolution Reveals a Universe without Design*. New York: Norton.

———. 1995. *River Out of Eden: A Darwinian View of Life*. New York: Basic Books.

———. 1998. *Unweaving the Rainbow*. Boston: Houghton Mifflin.

———. 2004. *The Ancestor's Tale: A Pilgrimage to the Dawn of Life*. Boston: Houghton Mifflin.

de Queiroz, K. 1999. The General Lineage Concept of Species and the Defining Properties of the Species Category. In R. A. Wilson, (ed.), *Species: New Interdisciplinary Essays*. Cambridge, MA: MIT Press.

de Vries, H. 1906. *Species and Varieties: Their Origin by Mutation*. Chicago: Open Court.

———. 1909. Variation. In A. C. Seward, ed., *Darwin and Modern Science*. Cambridge: Cambridge University Press.

Deacon, T. W. 1998. *The Symbolic Species: The Co-evolution of Language and the Brain*. New York: Norton.

Dennett, D. C. 1974. Why the Law of Effect Won't Go Away. Reprinted in *Brainstorms: Philosophical Essays on Mind and Psychology*. Cambridge, MA: MIT Press.

———. 1981. *Brainstorms: Philosophical Essays on Mind and Psychology*. Cambridge, MA: MIT Press.

———. 1987. *The Intentional Stance*. Cambridge, MA: MIT Press.

———. 1995. *Darwin's Dangerous Idea: Evolution and the Meanings of Life*. New York: Simon & Schuster.

Devitt, M. 2008. Resurrecting Biological Essentialism. *Philosophy of Science* 75: 344–82.

Diderot, D. 1749/1992. Letter on the Blind. In J. H. Mason and R. Wokler, eds., *Diderot: Political Writings*. Cambridge: Cambridge University Press.

Djebali, S., Davis, C., et al. 2012. Landscape of Transcription in Human Cells. *Nature* 489: 101–8.

Dobzhansky, T. 1955. *Evolution, Genetics, and Man*. New York: Wiley.

Doebeli, M. and Ispolatov, I. 2010. A Model for the Evolutionary Diversification of Religions. *Journal of Theoretical Biology* 267: 676–84.

Donaldson-Matasci, M. C., Bergstrom, C. T., and Lachmann, M. 2010. The Fitness Value of Information. *OIKOS* 119: 219–30.

Doolittle, W. F. and Bapteste, E. 2007. Pattern Pluralism and the Tree of Life Hypothesis. *Proceeding of the National Academy of Sciences* 104: 2043–49.

Downes, S. M. 2011. Scientific Models. *Philosophy Compass* 6: 757–64.

Downes, S. M. and Machery, E., eds. 2013. *Arguing about Human Nature: Contemporary Debates*. London: Routledge.

Dretske, F. I. 1981. *Knowledge and the Flow of Information*. Cambridge, MA: MIT Press.

———. 1988. *Explaining Behavior: Reasons in a World of Causes*. Cambridge, MA: MIT Press.

Driesch, H. 1914. *The History and Theory of Vitalism*. London: Hesperides Press.

Dupré, J. 1993. *The Disorder of Things: Metaphysical Foundations of the Disunity of Science*. Cambridge, MA: Harvard University Press.

———. 1999. On the Impossibility of a Monistic Account of Species. In R. A. Wilson, ed., *Species: New Interdisciplinary Essays*. Cambridge, MA: MIT Press.

———. 2001. *Human Nature and the Limits of Science*. New York: Oxford University Press.

———. 2006. *Humans and Other Animals*. New York: Oxford University Press.

———. 2012. *Processes of Life: Essays in the Philosophy of Biology*. Oxford: Oxford University Press.

Dupré, J. and O'Malley, M. A. 2009. Varieties of Living Things: Life at the Intersection of Lineage and Metabolism. *Philosophy and Theory in Biology* 1: 1–24.

Durham, W. H. 1992. *Coevolution: Genes, Culture, and Human Diversity*. Palo Alto: Stanford University Press.

Edelman, G. M. 1987. *Neural Darwinism: The Theory of Neuronal Group Selection*. New York: Basic Books.

Edgington, D. 2008. Counterfactuals. *Proceedings of the Aristotelian Society* 108: 1–21.

Ehrlich, P. R. and Raven, P. H. 1969. Differentiation of Populations. *Science* 165: 1228–33.

Ereshefsky, M. 1992. Eliminative Pluralism. *Philosophy of Science* 59: 671–90.

———. 1998. Species Pluralism and Anti-Realism. *Philosophy of Science* 65: 103–20.

———. 2010. Microbiology and the Species Problem. *Biology and Philosophy* 25: 553–68.

Falk, R. 2009. *Genetic Analysis: A History of Genetic Thinking*. Cambridge: Cambridge University Press.

Fisher, R. A. 1930. *The Genetical Theory of Natural Selection*. Oxford: Clarendon.

Folse, H. J. and Roughgarden, J. 2010. What Is an Individual Organism? A Multilevel Selection Perspective. *Quarterly Review of Biology* 85: 447–72.

Forber, P. 2005. On the Explanatory Roles of Natural Selection. *Biology and Philosophy* 20: 329–42.

Frank, S. A. 2012. Natural Selection. IV. The Price Equation. *Journal of Evolutionary Biology* 25: 1002–19.

Franklin, L. R. 2007. Bacteria, Sex, and Systematics. *Philosophy of Science* 74: 69–95.

Franklin-Hall, L. R. 2010. Trashing Life's Tree. *Biology and Philosophy* 25: 689–709.

Frigg, R. 2010. Models and Fiction. *Synthese* 172: 251–68.

Gallistel, C. R. and King, A. P. 2010. *Memory and the Computational Brain: Why Cognitive Science Will Transform Neuroscience*. New York: Wiley-Blackwell.

Gardner, A. and Welch, J. J. 2011. A Formal Theory of the Selfish Gene. *Journal of Evolutionary Biology* 24: 1801–13.

Gerstein, M. B., Can, B., Rozowsky, J., Deyou, Z., Du, J., Korbel, J., Emanuelsson, O., Zhang, Z., Weissman, S., and Snyder, M. 2007. What Is a Gene, post-ENCODE? History and Updated Definition. *Genome Research* 17: 669–81.

Ghiselin, M. 1969. *The Triumph of the Darwinian Method*. Berkeley: University of California Press.

———. 1974. A Radical Solution to the Species Problem. *Systematic Zoology* 23: 536–44.

———. 1997. *Metaphysics and the Origin of Species*. Albany: State University of New York Press.

Giere, R. N. 1988. *Explaining Science: A Cognitive Approach*. Chicago: University of Chicago Press.

———. 1999. *Science Without Laws*. Chicago: University of Chicago Press.

Gillespie, J. 1977. Natural Selection for Variances in Offspring Numbers— A New Evolutionary Principle. *American Naturalist* 111: 1010–14.

Ginzburg, L. and Colyvan, M. 2004. *Ecological Orbits: How Planets Move and Populations Grow*. New York. Oxford University Press.

Gleick, J. 2011. *The Information: A History, a Theory, a Flood*. New York: Pantheon.

Glennan, S. 1996. Mechanisms and the Nature of Causation. *Erkenntnis* 44: 49–71.

———. 2002. Rethinking Mechanistic Explanation. *Philosophy of Science* 69: S342–53.

Godfrey-Smith, P. 1994. A Modern History Theory of Functions. *Noûs* 28: 344–62.

———. 1996. *Complexity and the Function of Mind in Nature*. Cambridge: Cambridge University Press.

———. 2000. On the Theoretical Role of "Genetic Coding." *Philosophy of Science* 67: 26–44.

———. 2001a. Organism, Environment and Dialectics. In R. Singh, C. Krimbas, D. Paul, and J. Beatty, eds., *Thinking about Evolution: Historical, Philosophical, and Political Perspectives*. Cambridge: Cambridge University Press.

———. 2001b. Three Kinds of Adaptationism. In S. H. Orzack and E. Sober, eds., *Adaptationism and Optimality*. Cambridge: Cambridge University Press.

———. 2006. The Strategy of Model-Based Science. *Biology and Philosophy* 21: 725–40.

———. 2009. *Darwinian Populations and Natural Selection*. Oxford: Oxford University Press.

———. 2010. Causal Pluralism. In C. Hitchcock and P. Menzies, eds., *Oxford Handbook of Causation*. Oxford: Oxford University Press.

———. 2012. Darwinism and Cultural Change. *Philosophical Transactions of the Royal Society B* 367: 2160–70.

———. 2013. Darwinian Individuals. In F. Bouchard and P. Huneman, eds., *From Groups to Individuals: Evolution and Emerging Individuality*. Cambridge, MA: MIT Press.

Godfrey-Smith, P. and Kerr, B. 2009. Selection in Ephemeral Networks. *American Naturalist* 174: 906–11.

Gould, S. J. 1980. Caring Groups and Selfish Genes. In *The Panda's Thumb: More Reflections in Natural History*. New York: Norton.

———. 2002. *The Structure of Evolutionary Theory*. Cambridge, MA: Belknap.

Gould, S. J. and Lewontin, R. C. 1979. The Spandrels of San Marco and the Panglossian Paradigm: A Critique of the Adaptationist

Programme. *Proceedings of the Royal Society of London, Series B* 205: 581–98.

Gray, R. D., Greenhill, S. J., and Ross, R. M. 2007. The Pleasures and Perils of Darwinizing Culture (with Phylogenies). *Biological Theory* 2: 360–75.

Griesemer, J. 2000. The Units of Evolutionary Transition. *Selection* 1: 67–80.

———. 2005. The Informational Gene and the Substantial Body: On the Generalization of Evolutionary Theory by Abstraction. In M. Jones and N. Cartwright, eds., *Idealization XII: Correcting the Model, Idealization and Abstraction in the Sciences.* Amsterdam: Rodopi.

Griffiths, P. E. 1999. Squaring the Circle: Natural Kinds with Historical Essences. In R. A. Wilson, ed., *Species: New Interdisciplinary Essays.* Cambridge, MA: MIT Press.

———. 2001. Genetic Information: A Metaphor in Search of a Theory. *Philosophy of Science* 68: 394–412.

———. 2002. What Is Innateness? *The Monist* 85: 70–85.

Griffiths, P. E. and Gray, R. D. 1994. Developmental Systems and Evolutionary Explanation. *Journal of Philosophy* 91: 277–304.

Griffiths, P. E. and Stotz, K. 2007. Gene. In D. Hull and M. Ruse, eds., *Cambridge Companion for the Philosophy of Biology.* Cambridge: Cambridge University Press.

———. 2013. *Genetics and Philosophy: An Introduction.* Cambridge: Cambridge University Press.

Güth, W., Schmittberger, R., and Schwarze, B. 1982. An Experimental Analysis of Ultimatum Bargaining. *Journal of Economic Behavior and Organization* 3: 367–88.

Haig, D. 1997. The Social Gene. In J. R. Krebs and N. B. Davies, eds., *Behavioural Ecology: An Evolutionary Approach.* 4th ed. London: Blackwell.

———. 2012. The Strategic Gene. *Biology and Philosophy* 27: 461–79.

Haldane, J. B. S. 1932. *The Causes of Evolution.* London: Longmans, Green.

Hamilton, W. D. (1964) The Genetical Evolution of Social Behaviour I, II. *Journal of Theoretical Biology* 7: 1–16, 7–64.

———. 1975. Innate Social Aptitudes of Man: An Approach from Evolutionary Genetics. In R. Fox, ed., *Biosocial Anthropology.* London: Malaby Press.

———. 1998. *Narrow Roads of Gene Land Vol. 1: Evolution of Social Behavior.* Oxford: Oxford University Press.

Harman, O. 2011. *The Price of Altruism: George Price and the Search for the Origins of Kindness.* New York: Norton.

Harms, W. 2004. Primitive Content, Translation, and the Emergence of Meaning in Animal Communication. In D. K. Oller and U. Griebel (eds.), *Evolution of Communication Systems: A Comparative Approach*. Cambridge: MIT Press.

Harper, J. 1977. *Population Biology of Plants*. Caldwell, NJ: Blackburn Press.

Hempel, C. 1966. *Philosophy of Natural Science*. Englewood Cliffs, NJ: Prentice Hall.

Hennig, W. 1966. *Phylogenetic Systematics*. Translated by D. David and R. Zangerl. Urbana: University of Illinois Press.

Henrich, J., Boyd, R., Bowles, S., Camerer, C., Fehr, E., and Gintis, H., eds. 2004. *Foundations of Human Sociality: Economic Experiments and Ethnographic Evidence from Fifteen Small-Scale Societies*. New York: Oxford University Press.

Hey, J. 2011. Regarding the Confusion between the Population Concept and Mayr's "Population Thinking." *Quarterly Review of Biology* 86: 253–64.

Heyes, C. 2012. Grist and Mills: On the Cultural Origins of Cultural Learning. *Philosophical Transactions of the Royal Society of London B, Biological Sciences* 367: 2181–91.

Hodgson, G. M. and Knudsen, T. 2010. *Darwin's Conjecture: The Search for General Principles of Social and Economic Evolution*. Chicago: University of Chicago Press.

Hölldobler, B. and Wilson, E. O. 2008. *The Superorganism: The Beauty, Elegance, and Strangeness of Insect Societies*. New York: Norton.

Horwich, P. 1990. *Truth*. New York: Oxford University Press.

Hull, D. L. 1970. Contemporary Systematic Philosophies. *Annual Review of Ecology and Systematics* 1: 19–54.

———. 1976. Are Species Really Individuals? *Systematic Zoology* 25: 174–91.

———. 1980. Individuality and Selection. *Annual Review of Ecology and Systematics* 11: 311–32.

———. 1986. On Human Nature. *PSA: Proceedings of the Biennial Meeting of the Philosophy of Science Association* 2: 3–13.

———. 1988. *Science as a Process: An Evolutionary Account of the Social and Conceptual Development of Science*. Chicago: University of Chicago Press.

Hull, D. L., Langman, R. E., and Glenn, S. S. 2001. A General Analysis of Selection. *Behavioral and Brain Sciences* 24: 511–73.

Hume, D. 1739. *A Treatise of Human Nature: Being an Attempt to Introduce the Experimental Method of Reasoning into Moral Subjects*. London: J. Noon.

Huneman, P., ed. 2012. *Functions: Selection and Mechanisms*. Dordrecht: Springer.

References

Huxley, J. 1912. *The Individual in the Animal Kingdom*. Cambridge: Cambridge University Press.

———. 1942. *Evolution: The Modern Synthesis*. London: Allen & Unwin.

Huxley, T. H. 1852. Upon Animal Individuality. *Proceedings of the Royal Institution of Great Britain* 1: 184–89.

Istrail, S., De-Leon, S., and Davison, E. 2007. The Regulatory Genome and the Computer. *Developmental Biology* 310: 187–95.

Jablonka, E. and Lamb, M. 2005. *Evolution in Four Dimensions: Genetic, Epigenetic, Behavioral, and Symbolic Variation in the History of Life*. Cambridge, MA: MIT Press.

Jackson, J., Buss, L., and Cook, R., eds. 1985. *Population Biology and Evolution of Clonal Organisms*. New Haven: Yale University Press.

Jacob, F. and Monod, J. 1961. Genetic Regulatory Mechanisms in the Synthesis of Proteins. *Journal of Molecular Biology* 3: 318–56.

James, W. 1907. *Pragmatism, a New Name for Some Old Ways of Thinking: Popular Lectures on Philosophy*. New York: Longmans, Green.

Janzen, D. H. 1977. What Are Dandelions and Aphids? *American Naturalist* 111: 586–89.

Jerne, N. K. 1955. The Natural Selection Theory of Antibody Formation. *Proceedings of the National Academy of Sciences USA* 41: 849–57.

Johannsen, W. 1909. *Elemente der exakten erblichkeitslehre*. Jena: Gustav Fischer.

Jones, C. G., Lawton, J. H., and Shachak, M. 1994. Organisms as Ecosystem Engineers. *Oikos* 69: 373–86.

Judson, H. F. 1996. *The Eighth Day of Creation: Makers of the Revolution in Biology*. 25th Anniversary ed. Cold Spring Harbor, NY: Cold Spring Harbor Laboratory Press.

Kant, I. 1790/1987. *Critique of Judgment*. Translated by W. S. Pluhar. Indianapolis: Hackett.

Kay, L. 2000. *Who Wrote the Book of Life? A History of the Genetic Code*. Palo Alto: Stanford University Press.

Kerr, B. and Godfrey-Smith, P. 2002. Individualist and Multi-Level Perspectives on Selection in Structured Populations. *Biology and Philosophy* 17: 477–517.

Kerr, B., Godfrey-Smith, P., and Feldman, M. 2004. What Is Altruism? *Trends in Ecology and Evolution* 19: 135–40.

Kingsland, S. E. 1995. *Modeling Nature: Episodes in the History of Population Ecology*. Chicago: University of Chicago Press.

Kirk, D. L. 1998. *Volvox: Molecular-Genetic Origins of Multicellularity and Cellular Differentiation*. Cambridge: Cambridge University Press.

Kitcher, P. 1984. Species. *Philosophy of Science* 51: 308–33.

———. 2011. *The Ethical Project.* Cambridge, MA: Harvard University Press.

Koch, C. 1999. *Biophysics of Computation: Information Processing in Single Neurons.* New York: Oxford University Press.

Krugman, P. 2009. How Did Economists Get It So Wrong? *New York Times*, September 2.

Kuhn, S. 2009. Prisoner's Dilemma. In E. N. Zalta, ed., *Stanford Encyclopedia of Philosophy.* Spring 2009 ed. http://plato.stanford.edu/archives/spr2009/entries/prisoner-dilemma/.

Lamarck, J.-B. 1809/2011. *Zoological Philosophy: An Exposition with Regard to the Natural History of Animals.* Translated by H. S. R. Elliott. Cambridge: Cambridge University Press.

Langton, R. and Lewis, D. 1998. Defining "Intrinsic." *Philosophy and Phenomenological Research* 58: 333–45.

Laublichler, M. and Maienschein, J., eds. 2009. *From Embryology to Evo-Devo: A History of Developmental Evolution.* Cambridge, MA: MIT Press.

Lem, S. 1961/1970. *Solaris.* Translated by J. Kilmartin and S. Cox. New York: Walker.

Levins, R. 1966. The Strategy of Model Building in Population Biology. *American Scientist* 54: 421–31.

———. 1970. Complexity. In C. H. Waddington, ed., *Towards a Theoretical Biology.* Vol. 3. Edinburgh: University of Edinburgh Press.

Levins, R. and Lewontin, R. C. 1985. *The Dialectical Biologist.* Cambridge, MA: Harvard University Press.

Levy, A. 2010. Information in Biology: A Fictionalist Account. *Noûs* 45: 640–57.

———. 2013. Three Kinds of New Mechanism. *Biology and Philosophy* 28: 99–114.

Lewens, T. 2006. *Darwin.* London: Routledge.

———. 2009. What's Wrong with Typological Thinking? *Philosophy of Science* 79: 355–71.

———. 2010. Natural Selection Then and Now. *Biological Reviews* 85: 829–35.

———. 2012. Pheneticism Reconsidered. *Biology and Philosophy* 27: 159–77.

Lewis, D. K. 1969. *Convention: A Philosophical Study.* Cambridge, MA: Harvard University Press.

———. 2000. Causation as Influence. *Journal of Philosophy* 97: 182–97.

Lewontin, R. C. 1970. The Units of Selection. *Annual Review of Ecology and Systematics* 1: 1–18.

———. 1981. *Biology as Ideology: The Doctrine of DNA.* New York: Harper Perennial.

―――. 1983. The Organism as the Subject and Object of Evolution. In R. Levins and R. C. Lewontin, eds., *The Dialectical Biologist*. Cambridge, MA: Harvard University Press.

―――. 1985. Adaptation. In R. Levins and R. C. Lewontin, eds., *The Dialectical Biologist*. Cambridge, MA: Harvard University Press.

Lewontin, R. C. 1991. *Biology as Ideology: The Doctrine of DNA*. New York: HarperCollins.

Linnaeus, C. 1758. *Systema Naturæ Per Regna Tria Naturæ, Secundum Classes, Ordines, Genera, Species, Cum Characteribus, Differentiis, Synonymis, Locis*. 10th ed. Stockholm: Laurentius Salvius.

Lloyd, E. A. 1988. *The Structure and Confirmation of Evolutionary Theory*. Westport: Greenwood.

―――. 2001. Units and Levels of Selection: An Anatomy of the Units of Selection Debates. In R. Singh, C. Krimbas, D. Paul, and J. Beatty, eds., *Thinking about Evolution: Historical, Philosophical, and Political Perspectives*. Cambridge: Cambridge University Press.

Lovejoy, A. O. 1936. *The Great Chain of Being: A Study of the History of an Idea*. Cambridge, MA: Harvard University Press.

Lovelock, J. 2000. *Gaia: A New Look at Life on Earth*. 3rd ed. New York: Oxford University Press.

Lyell, C. 1830. *Principles of Geology*. London: John Murray.

Machamer, P., Darden, L., and Craver, C. 2000. Thinking about Mechanisms. *Philosophy of Science* 67: 1–25.

Machery, E. 2008. A Plea for Human Nature. *Philosophical Psychology* 21: 321–30.

Mallet, J. 1995. A Species Definition for the Modern Synthesis. *Trends in Ecology and Evolution* 10: 294–99.

Malthus, T. R. 1798. *An Essay on the Principle of Population*. London: J. Johnson.

Matthewson, J. and Calcott, B. 2011. Mechanistic Models of Population-Level Phenomena. *Biology and Philosophy* 26: 737–56.

Matthewson, J. and Weisberg, M. 2009. The Structure of Tradeoffs in Model Building. *Synthese* 170: 169–90.

Maynard Smith, J. 1982. *Evolution and the Theory of Games*. Cambridge: Cambridge University Press.

―――. 1987. How to Model Evolution. In J. Dupré, ed., *The Latest on the Best: Essays on Evolution and Optimality*. Cambridge, MA: MIT Press.

―――. 2000. The Concept of Information in Biology. *Philosophy of Science* 67: 177–94.

Maynard Smith, J. and Price, G. R. 1973. The Logic of Animal Conflict. *Nature* 246: 15–18.

Maynard Smith, J. and Szathmáry, E. 1995. *The Major Transitions in Evolution*. New York: Oxford University Press.

Mayr, E. 1942. *Systematics and the Origin of Species from the Viewpoint of a Zoologist*. Cambridge, MA: Harvard University Press.

———. 1959. Darwin and the Evolutionary Theory in Biology. In B. J. Meggers, ed., *Evolution and Anthropology: A Centennial Appraisal*. Washington, DC: Anthropological Society of Washington.

———. 1969. *Principles of Systematic Zoology*. New York: McGraw-Hill.

McOuat, G. 2009. The Origins of "Natural Kinds": Keeping "Essentialism" at Bay in the Age of Reform. *Intellectual History Review* 19: 211–30.

McLaughlin, B. 1992. The Rise and Fall of British Emergentism. In A. Beckermann, H. Flohr, and J. Kim, eds., *Emergence or Reduction?* Berlin: Walter de Gruyter.

McShea, D. 1991. Complexity and Evolution: What Everybody Knows. *Biology and Philosophy* 6: 303–24.

McShea, D. W. and Brandon, R. N. 2010. *Biology's First Law: The Tendency for Diversity and Complexity to Increase in Evolutionary Systems*. Chicago: University of Chicago Press.

Medin, D. and Atran, S., eds. 1999. *Folkbiology*. Cambridge: Bradford Books.

Meehan, C. J., Olson, E. J., Reudink, M. W., Kyser, T. K., and Curry, R. L. 2009. Herbivory in a Spider through Exploitation of an Ant-Plant Mutualism. *Current Biology* 19: 892–93.

Mesoudi, A., Whiten, A., and Laland, K. N. 2006. Towards a Unified Science of Cultural Evolution. *Behavioral and Brain Sciences* 29: 329–83.

Michod, R. E. 1999. *Darwinian Dynamics: Evolutionary Transitions in Fitness and Individuality*. Princeton: Princeton University Press.

Michod, R. E. and Sanderson, M. 1985. Behavioural Structure and the Evolution of Social Behavior. In P. J. Greenwood and M. Slatkin, eds., *Evolution—Essays in Honour of John Maynard Smith*. Cambridge: Cambridge University Press.

Millikan, R. G. 1984. *Language, Thought and Other Biological Categories: New Foundations for Realism*. Cambridge, MA: MIT Press.

———. 2004. *Varieties of Meaning: The 2002 Jean Nicod Lectures*. Cambridge, MA: MIT Press.

Mills, S. K. and Beatty, J. H. 1979. The Propensity Interpretation of Fitness. *Philosophy of Science* 46: 263–386.

Millstein, R. L. 2006. Natural Selection as a Population-Level Causal Process. *British Journal for the Philosophy of Science* 574: 627–53.

Mishler, B. D. 1999. Getting Rid of Species. In R. A. Wilson, ed., *Species: New Interdisciplinary Essays*. Cambridge, MA: MIT Press.

Mitchell, S. 2000. Dimensions of Scientific Law. *Philosophy of Science* 67: 242–65.

Mitton, J. B. and Grant, M. C. 1996. Genetic Variation and the Natural History of Quaking Aspen. *BioScience* 46: 25–31.

Moss, L. 2002. *What Genes Can't Do*. Cambridge, MA: MIT Press.

Nagel, T. 2012. *Mind and Cosmos: Why the Materialist New-Darwinian Conception of Nature Is Almost Certainly False*. New York: Oxford University Press.

Nanney, D. L. 1958. Epigenetic Control Systems. *Proceedings of the National Academy of Sciences USA* 44: 712–17.

Neander, K. 1991. The Teleological Notion of "Function." *Australasian Journal of Philosophy* 69: 454–68.

———. 1995. Pruning the Tree of Life. *British Journal for the Philosophy of Science* 46: 59–80.

Nowak, M. A. 2006a. *Evolutionary Dynamics: Exploring the Equations of Life*. Cambridge, MA: Harvard University Press.

———. 2006b. Five Rules for the Evolution of Cooperation. *Science* 314: 1560–63.

O'Hara, R. J. 1991. Representations of the Natural System in the Nineteenth Century. *Biology and Philosophy* 6: 255–74.

Odling-Smee, F. J., Laland, K. N., and Feldman, M. W. 2003. *Niche Construction: The Neglected Process in Evolution*. Princeton: Princeton University Press.

Okasha, S. 2002. Darwinian Metaphysics: Species and the Question of Essentialism. *Synthese* 131: 191–213.

———. 2006. *Evolution and the Levels of Selection*. New York: Oxford University Press.

Orzack, S. and Sober, E., eds. 2001. *Adaptationism and Optimality*. Cambridge: Cambridge University Press.

Oyama, S. 1985. *The Ontogeny of Information: Developmental Systems and Evolution*. Cambridge: Cambridge University Press.

———. 2000. *Evolution's Eye: A Systems View of the Biology-Culture Divide*. Durham: Duke University Press.

Oyama, S., Griffiths, P. E., and Gray, R. D., eds. 2001. *Cycles of Contingency: Developmental Systems and Evolution*. Cambridge, MA: MIT Press.

Paley, W. 1802/2006. *Natural Theology; Or, Evidences of the Existence and Attributes of the Deity*. New York: Oxford University Press.

Pearce, T. forthcoming. The Dialectical Biologist, circa 1890: John Dewey and the Oxford Hegelians. *Journal of the History of Philosophy*.

Pearl, J. 2000. *Causality: Models, Reasoning, and Inference*. Cambridge: Cambridge University Press.

Pepper, J. W. and Herron, M. D. 2008. Does Biology Need an Organism Concept? *Biological Reviews* 83: 621–27.

Pigliucci, M. and Kaplan, J. 2003. On the Concept of Biological Race and Its Applicability to Humans. *Philosophy of Science* 70: 1161–72.

Pinker, S. 2002. *The Blank Slate: The Modern Denial of Human Nature.* New York: Viking.

Pittendrigh, C. S. 1958. Adaptation, Natural Selection, and Behavior. In A. Roe and G. G. Simpson, eds., *Behavior and Evolution.* New Haven: Yale University Press.

Popper, K. 1959. *The Logic of Scientific Discovery.* London: Routledge.

Portin, P. 2002. Historical Development of the Concept of the Gene. *Journal of Medicine and Philosophy* 27: 257–86.

Pradeu, T. 2010. What Is an Organism? An Immunological Answer. *History and Philosophy of the Life Sciences* 32: 247–68.

———. 2012. *The Limits of the Self: Immunology and Biological Identity.* New York: Oxford University Press.

Price, G. R. 1970. Selection and Covariance. *Nature* 227: 520–21.

———. 1972. Extension of Covariance Selection Mathematics. *Annals of Human Genetics* 35: 485–90.

Prinz, J. 2012. *Beyond Human Nature: How Culture and Experience Shape the Human Mind.* New York: Norton.

Provine, W. B. 1971. *The Origins of Theoretical Population Genetics.* Chicago: University of Chicago Press.

Ptashne, M. and Gann, A. 2001. *Genes and Signals.* Cold Spring Harbor, NY: Cold Spring Harbor Laboratory Press.

Puzo, M. and Coppola, F. F. 1972. *The Godfather.* Screenplay.

Queller, D. C. 1985. Kinship, Reciprocity and Synergism in the Evolution of Social Behaviour. *Nature* 318: 366–67.

———. 1997. Cooperators since Life Began (A Review of *The Major Transitions in Evolution,* by J. Maynard Smith and E. Szathmary). *Quarterly Review of Biology* 72: 184–88.

Queller, D. C. and Strassman, J. 2009. Beyond Society: The Evolution of Organismality. *Philosophical Transactions of the Royal Society B* 364: 3143–55.

Richerson, P. J. and Boyd, R. 2005. *Not by Genes Alone: How Culture Transformed Human Evolution.* Chicago: University of Chicago Press.

Ridley, M. 2007. *Evolution.* 3rd ed. New York: Wiley.

Rogers, D. S. and Ehrlich, P. R. 2008. Natural Selection and Cultural Rates of Change. *Proceedings of the National Academy of Sciences USA.* 105: 3416–20.

Rosenberg, A. and Kaplan, D. M. 2005. How to Reconcile Physicalism and Antireductionism about Biology. *Philosophy of Science* 72: 43–68.

Santelices, B. 1999. How Many Kinds of Individuals Are There? *Trends in Ecology and Evolution* 14: 152–55.

Sartre, J.-P. 1946/1956. Existentialism Is a Humanism. Reprinted in W. Kaufman, ed., *Existentialism from Dostoyevsky to Sartre*. New York: Penguin.

Schlichting, C. D. and Pigliucci, M. 1998. *Phenotypic Evolution: A Reaction Norm Perspective*. Sunderland: Sinauer Associates.

Schmalhausen, I. I. 1949. *Factors of Evolution: The Theory of Stabilizing Selection*. Philadelphia: Blankston.

Schweber, S. 1977. The Origin of the "Origin" Revisited. *Journal of the History of Biology* 10: 229–316.

Seabright, P. 2010. *The Company of Strangers: A Natural History of Economic Life*. Rev. ed. Princeton: Princeton University Press.

Searcy, W. and Nowicki, S. 2006. *The Evolution of Animal Communication*. Princeton: Princeton University Press.

Sellars, W. 1962. Philosophy and the Scientific Image of Man. In R. Colodny, ed., *Frontiers of Science and Philosophy*. Pittsburgh: University of Pittsburgh Press.

Shannon, C. E. 1948. A Mathematical Theory of Communication. *Bell System Technical Journal* 27: 379–423, 623–56.

Shea, N. 2007. Representation in the Genome, and in Other Inheritance Systems. *Biology and Philosophy* 22: 313–31.

———. 2012. Two Modes of Transgenerational Information Transmission. In K. Sterelny, R. Joyce, B. Calcott, and B. Fraser, eds., *Cooperation and Its Evolution*. Cambridge, MA: MIT Press.

Skinner, B. F. 1974. *About Behaviorism*. New York: Knopf Doubleday.

Skipper, R. A. and Millstein, R. L. 2005. Thinking about Evolutionary Mechanisms: Natural Selection. *Studies in History and Philosophy of Science Part C* 36: 327–47.

Skrupskelis, I. 2007. Evolution and Pragmatism: An Unpublished Letter of William James. *Transactions of the Charles S. Peirce Society* 43: 745–52.

Skyrms, B. 1980. *Causal Necessity*. New Haven: Yale University Press.

———. 1996. *Evolution of the Social Contract*. Cambridge: Cambridge University Press.

———. 2004. *The Stag Hunt and the Evolution of Social Structure*. Cambridge: Cambridge University Press.

———. 2010. *Signals: Evolution, Learning, and Information*. New York: Oxford University Press.

Smart, J. J. C. 1959. Can Biology Be an Exact Science? *Synthese* 11: 359–68.

Smith, A. 1776. *The Wealth of Nations*. London: W. Strahan and T. Cadell.

Sober, E. 1980. Evolution, Population Thinking, and Essentialism. *Philosophy of Science* 47: 350–83.

———. 1984. *The Nature of Selection: Evolutionary Theory in Philosophical Focus*, Cambridge: Bradford Books.

———. 1993. *Philosophy of Biology*. Boulder: Westview Press.

———. 1997. Two Outbreaks of Lawlessness in Recent Philosophy of Biology. *Philosophy of Science* 64: S458–67.

———. 2001. The Two Faces of Fitness. In R. Singh, C. Krimbas, D. Paul, and J. Beatty, eds., *Thinking about Evolution: Historical, Philosophical, and Political Perspectives*. Cambridge: Cambridge University Press.

———. 2011. *Did Darwin Write the Origin Backwards?* New York: Prometheus Books.

Sober, E. and Lewontin, R. C. 1982. Artifact, Cause and Genic Selection. *Philosophy of Science* 49: 157–80.

Sober, E. and Wilson, D. S. 1998. *Unto Others: The Evolution and Psychology of Unselfish Behavior*. Cambridge, MA: Harvard University Press.

Sokal, R. and Sneath, P. 1963. *Principles of Numerical Taxonomy*. San Francisco: W. H. Freeman.

Spencer, H. 1864. *Principles of Biology*. Vol. 1. London: Williams and Norgate.

Sperber, D. 1996. *Explaining Culture: A Naturalistic Approach*. Oxford: Blackwell.

———. 2000. An Objection to the Memetic Approach to Culture. In R. Aunger, ed., *Darwinizing Culture: The Status of Memetics*. New York: Oxford University Press.

Stegmann, U., ed. 2013. *Animal Communication Theory: Information and Influence*. Cambridge: Cambridge University Press.

Sterelny, K. 2003. *Thought in a Hostile World: The Evolution of Human Cognition* Oxford: Blackwell.

———. 2011. Darwinian Spaces: Peter Godfrey-Smith on Selection and Evolution. *Biology and Philosophy* 26: 489–500.

———. 2012. *The Evolved Apprentice: How Evolution Made Humans Unique* Cambridge, MA: MIT Press.

———. Forthcoming. Cooperation, Culture and Conflict. *British Journal for the Philosophy of Science*.

Sterelny, K. and Griffiths, P. E. 1998. *Sex and Death: An Introduction to Philosophy of Biology*. Chicago: University of Chicago Press.

Sterelny, K. and Kitcher, P. S. 1989. The Return of the Gene. *Journal of Philosophy* 85: 339–61.

Sterelny, K., Smith, K. C., and Dickison, M. 1996. The Extended Replicator. *Biology and Philosophy* 11: 377–403.

References

Stott, R. 2012. *Darwin's Ghosts: The Secret History of Evolution*. New York: Spiegel & Grau.

Taylor, P. D. and Jonker, L. 1978. Evolutionary Stable Strategies and Game Dynamics. *Mathematical Biosciences* 40: 145–56.

Templeton, A. R. 1989. The Meaning of Species and Speciation: A Genetic Perspective. In D. Otte and J. A. Endler, eds., *Speciation and Its Consequences*. Sunderland, MA: Sinauer.

Thompson, M. 2004. Apprehending Human Form. *Royal Institute of Philosophy Supplement* 54: 47–74.

Thorndike, E. L. 1911. *Animal Intelligence*. New York: Macmillan.

Tomasello, M. 1999. *The Cultural Origins of Human Cognition*. Cambridge, MA: Harvard University Press.

———. 2008. *Origins of Human Communication*. Cambridge, MA: MIT Press.

———. 2009. *Why We Cooperate*. Cambridge, MA: MIT Press.

Toon, A. 2012. *Models as Make-Believe: Imagination, Fiction and Scientific Representation*. Houndmills: Palgrave Macmillan.

Trivers, R. L. 1971. The Evolution of Reciprocal Altruism. *Quarterly Review of Biology* 46: 35–57.

Turchin, P. 2001. Does Population Ecology Have General Laws? *OIKOS* 94: 17–26.

Turing, A. M. 1936. On Computable Numbers, with an Application to the Entscheidungsproblem. *Proceedings of the London Mathematical Society* 42: 230–65.

Ursell, L. K., Clemente, J. C., Rideout, J. R., Gevers, D., Caporaso, J. G., and Knight, R. 2012. The Interpersonal and Intrapersonal Diversity of Human-Associated Microbiota in Key Body Sites. *Journal of Allergy and Clinical Immunology* 129: 1204–8.

Van Valen, L. 1976. Ecological Species, Multispecies, and Oaks. *Taxon* 25: 233–39.

Velasco, J. 2012. The Future of Systematics: Tree-Thinking without the Tree. *Philosophy of Science* 79: 624–36.

Von Frisch, K. 1993. *The Dance Language and Orientation of Bees*. Translated by L. Chadwick. Cambridge, MA: Harvard University Press.

Waddington, C. H. 1942. Canalization of Development and the Inheritance of Acquired Characters. *Nature* 150: 563–65.

Wagner, G.P. Forthcoming. *Homology, Genes and Evolutionary Innovation*. (Princeton: Princeton University Press, forthcoming).

Wallace, A. R. 1858. On the Tendency of Varieties to Depart Indefinitely from the Original Type. *Proceedings of the Linnaean Society of London* 3: 53–62.

Walsh, D. M., Lewens, T., and Ariew, A. 2002. The Trials of Life: Natural Selection and Random Drift. *Philosophy of Science* 69: 452–73.

Waters, C. K. 1994. Genes Made Molecular. *Philosophy of Science* 61: 163–85.

———. 1998. Causal Regularities in the Biological World of Contingent Distributions. *Biology and Philosophy* 13: 5–36.

———. 2007. Causes That Make a Difference. *Journal of Philosophy* 104: 551–79.

Watson, J. D. and Crick, F. H. C. 1953. A Structure for Deoxyribose Nucleic Acid. *Nature* 171: 737–38.

Weatherson, B. 2006. Intrinsic vs. Extrinsic Properties. In E. N. Zalta, ed., *Stanford Encyclopedia of Philosophy*. Fall 2006 ed. http://plato.stanford.edu/archives/fall2006/entries/intrinsic-extrinsic/.

Weibull, J. 1995. *Evolutionary Game Theory*. Cambridge, MA: MIT Press.

Weisberg, M. 2007a. Three Kinds of Idealization. *Journal of Philosophy* 104: 639–59.

———. 2007b. Who Is a Modeler? *British Journal for Philosophy of Science* 58: 207–33.

———. 2013. *Simulation and Similarity: Using Models to Understand the World*. New York: Oxford University Press.

Weisberg, M. and Reisman, K. 2008. The Robust Volterra Principle. *Philosophy of Science* 75: 106–31.

West, G. B., Brown, J. H., and Enquist, B. J. 1997. A General Model for the Origin of Allometric Scaling Laws in Biology. *Science* 276: 122–26.

West, S. A., Griffin, A. S., and Gardner, A. 2007. Social Semantics: Altruism, Cooperation, Mutualism, Strong Reciprocity and Group Selection. *Journal of Evolutionary Biology* 20: 415–32.

West-Eberhard, M. J. 2003. *Developmental Plasticity and Evolution*. New York: Oxford University Press.

———. 2005. Developmental Plasticity and the Origin of Species Differences. *Proceedings of the National Academy of Sciences USA* 102: 6543–49.

Wilkins, J. S. 2009. *Species: A History of the Idea*. Berkeley: University of California Press.

Wilson, R. A., ed. 1999. *Species: New Interdisciplinary Essays*. Cambridge, MA: MIT Press.

Williams, G. C. 1966. *Adaptation and Natural Selection: A Critique of Some Current Evolutionary Thought*. Princeton: Princeton University Press.

———. 1992. *Natural Selection: Domains, Levels and Challenges*. New York: Oxford University Press.

Wilson, D. S. 2002. *Darwin's Cathedral: Evolution, Religion, and the Nature of Society*. Chicago: University of Chicago Press.

Wilson, E. O. and Bossert, W. H. 1971. *A Primer of Population Biology*. Sunderland: Sinauer Associates.

Wilson, R. A., Barker, M. J., and Brigandt, I. 2007. When Traditional Essentialism Fails. *Philosophical Topics* 35: 189–215.

Wimsatt, W. C. 1980. Reductionistic Research Strategies and Their Biases in the Units of Selection Controversy. In T. Nickles, ed., *Scientific Discovery Vol. II: Case Studies*. Dordrecht: Reidel.

———. 2006. Aggregate, Composed, and Evolved Systems: Reductionistic Heuristics as Means to More Holistic Theories. *Biology and Philosophy* 21: 667–702.

———. 2007. *Piecewise Approximations to Reality: Engineering a Philosophy of Science for Limited Beings*. Cambridge, MA: Harvard University Press.

Winsor, M. P. 2006. The Creation of the Essentialism Story: An Exercise in Metahistory. *History and Philosophy of the Life Sciences* 28: 149–74.

Woodward, J. 2001. Law and Explanation in Biology: Invariance Is the Kind of Stability That Matters. *Philosophy of Science* 68: 1–20.

———. 2003. *Making Things Happen: A Theory of Causal Explanation*. Oxford: Oxford University Press.

———. 2006. Sensitive and Insensitive Causation. *Philosophical Review* 115: 1–50.

Wrangham, R. 2010. *Catching Fire: How Cooking Made Us Human*. New York: Basic Books.

Wright, L. 1973. Functions. *Philosophical Review* 82: 139–68.

———. 1976. *Teleological Explanations: An Etiological Analysis of Goals and Functions*. Berkeley: University of California Press.

Wright, S. 1931. Evolution in Mendelian Populations. *Genetics* 16: 97–159.

———. 1932. The Roles of Mutation, Inbreeding, Crossbreeding and Selection in Evolution. *Proceedings of the Sixth International Congress of Genetics* 1: 356–66.

Index

Index

Churchill family, 109–11
cistrons, 83, 94
classification systems, 6, 116. *See also*
"kinds," biological; species
clockwork universe, 17
collective reproducers, 70–71
communication, 144, 146–56. *See also*
information
complexity, organized vs. disorganized, 51. *See also* organized
systems
computers, 22–23, 155
conditional statements: and different
kinds of generalizations in biology, 25; and economics, 26–27;
and laws of nature, 24–26; and
Lewontin's description of natural
selection, 30–32; material vs. subjunctive conditional statements,
26n; and natural selection, 33;
"other things being equal" (*ceteris
peribus* clause), 32; as products of
models, 22–26
consciousness, 18
cooking, 135
cooperation, 3, 120–31; defined/
described, 120; and evolutionary
transitions, 127–28; evolution
of, 43, 123–24, 127–31; and human societies, 131–36. *See also*
Prisoner's Dilemma game; *Stag
Hunt* game
Crick, Francis, 9, 82
culture: and adaptive change, 47, 48;
cultural evolution, 136–38; and
evolution of cooperation, 133

dandelions, 69, 71
Darwin, Charles: and natural selection, 29–33, 42; origin of ideas,
8; and social behavior, 121; and
species concept, 101; summary of
ideas, 7–8
Darwin, Erasmus, 6
Darwinian individuals, 68–79;
and cultural evolution, 136;

de-Darwinization of old entities,
75; distinguished from non-reproducing consortia, 78; distinguished
from organisms, 76–79; and human societies, 133–34; new types
arising from old types, 74–75
Dawkins, Richard: and evolution as
a flow of information, 145, 152;
and integrity of genes, 94, 99; and
selfish genes, 44; and universal
Darwinism, 47
Dennett, Daniel, 47, 136
The Descent of Man (Darwin), 121
determinacy, 64
developmental biology, 9, 10, 144
de Vries, Hugo, 38
Diderot, Denis, 6, 143
DNA, 9; and Central Dogma of
molecular biology, 14; cistrons,
83, 94; and evolution, 93–94; and
information, 145, 153–55; "junk"
DNA, 92; and memory, 153–55;
methylation, 155; and molecular
genetics, 82–83; noncoding DNA,
84; overlapping roles in cell,
88–89, 92–93, 98; and scaffolded
reproduction, 70. *See also* genes;
mutations
Dobzhansky, T., 51

E. coli bacteria, 20, 103
economics, 26–27, 63n
Einstein, Albert, 12
emergent properties, 18–19
Empedocles, 5, 41
empiricism, 54
environment: of alleles, 96; and genotype–phenotype relation, 87–88;
and learned traits, 141; organisms'
adaptations to, 51–55; organisms'
transformation/construction of,
54, 55–58
epigenetics, 155
Ereshefsky, Marc, 105, 107n, 112
ESS (evolutionarily stable strategy),
124

Milton Keynes UK
Ingram Content Group UK Ltd.
UKHW031641230924
448587UK00002B/15